THE SCIENCE OF
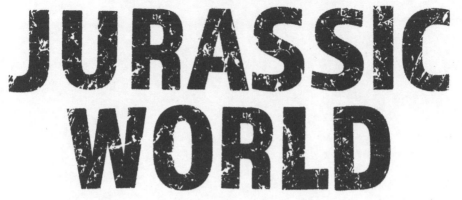
JURASSIC
WORLD

THE SCIENCE OF

JURASSIC WORLD

THE DINOSAUR FACTS BEHIND THE FILMS

MARK BRAKE & JON CHASE

AUTHORS OF *THE SCIENCE OF STAR WARS* AND *THE SCIENCE OF HARRY POTTER*

Skyhorse Publishing

Skyhorse Publishing books may be purchased in bulk at special discounts for sales promotion, corporate gifts, fund-raising, or educational purposes. Special editions can also be created to specifications. For details, contact the Special Sales Department, Skyhorse Publishing, 307 West 36th Street, 11th Floor, New York, NY 10018 or info@skyhorsepublishing.com.

Skyhorse® and Skyhorse Publishing® are registered trademarks of Skyhorse Publishing, Inc.®, a Delaware corporation.

Visit our website at www.skyhorsepublishing.com.

10 9 8 7 6 5 4 3 2 1

Library of Congress Cataloging-in-Publication Data is available on file.

Cover design by Daniel Brount
Cover photo by Getty Images

Print ISBN: 978-1-5107-6258-9
Ebook ISBN: 978-1-5107-6259-6

Printed in the United States of America

For Bryher, Eden, and Hallie

CONTENTS

DINOSAUR TIMELINE

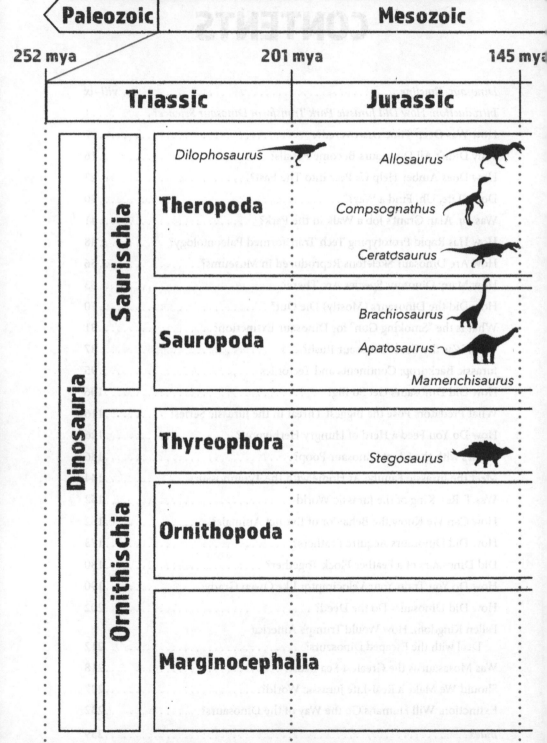

Paleozoic	Mesozoic	
252 mya	201 mya	145 mya
Triassic	Jurassic	

Dinosaur timeline showing all the dinosaurs of the movie franchise so far and when they existed

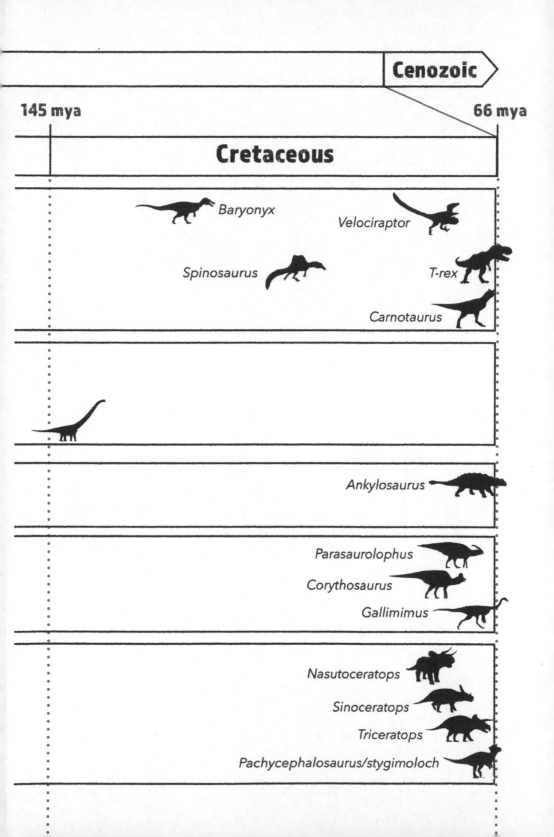

Cenozoic

145 mya   66 mya

Cretaceous

Baryonyx

Velociraptor

Spinosaurus

T-rex

Carnotaurus

Ankylosaurus

Parasaurolophus

Corythosaurus

Gallimimus

Nasutoceratops

Sinoceratops

Triceratops

Pachycephalosaurus/stygimoloch

INTRODUCTION: HOW DID JURASSIC PARK TRANSFORM DINOSAUR SCIENCE?

"I think the first *Jurassic Park* was the best thing that's ever happened to dinosaur paleontology. That led to an explosion of public interest in dinosaurs . . . and that led directly to a lot of museums putting out dinosaur exhibits. A lot of universities put out courses, and [there was] a lot more interest and money in the field. A lot of my colleagues got jobs specifically because of *Jurassic Park*, because a museum or university wanted to hire a paleontologist after that. So, I do think there is a really, really good chance I wouldn't have my job today if the book was never written, if the movie was never made. I think dinosaur paleontology right now would still be a really niche discipline, with only a handful of people around the world studying it."

—Interview with paleontologist Steve Brusatte,
The Verge (June 23, 2018)

"Until twenty years ago dinosaurs were widely assumed to be large lumpen lizards that became extinct millions of years ago. Discoveries in China have since shown dramatically that many were fast and feathered and some survived the great extinctions and are the ancestors of our modern birds. The recently discovered Chinese fossils of feathered dinosaurs are so well-preserved, scientists can even work out the feathers' color and where they were found on the dinosaurs' bodies and theorize about their use for displays, insulation and, in some cases perhaps, flight. Even a large T. rex may have had downy feathers, and it appears the small velociraptors had long quill-like feathers arranged on arms that look like wings."

—BBC Radio 4, "In Our Time: Feather Dinosaurs" (2017)

An Iconic Dinosaur Movie

Jurassic Park was not the first science fiction to influence science, of course. To name but a few early influences: The submarine was invented in the imagination of Jules Verne's 1870 book *Twenty Thousand Leagues Under the Sea*. The atomic bomb was conjured up by H. G. Wells in his 1914 novel *The World Set Free*. And NASA stole the 10-9-8-7-6 idea of the rocket countdown from Fritz Lang's 1929 sci-fi film *Frau im Mond* (*Woman in the Moon*). Over the years there have been many, many more.

But here's the unique thing about the science fiction of *Jurassic Park*: It influenced the way in which science looks at the past, rather than the future. The evolution of the field of dinosaur science, how it has blossomed by bounds and branched out in recent years, has been thanks in no small part to Steven Spielberg's iconic movie.

It all began in 1993. The year before had seen mostly mundane movies. But when *Jurassic Park* premiered on June 9 at the Uptown Theater in Washington, DC, and on general release from June 11 in the United States, it went on to gross over $914 million worldwide in its original theatrical run. *Jurassic Park* became the highest-grossing movie of that year, and the highest-grossing film ever at the time, a record held until the 1997 release of *Titanic*. Following *Jurassic Park*'s three-dimensional (3D) re-release in 2013, in celebration of its twentieth anniversary, the dinosaur movie became only the seventeenth film in history to surpass the $1 billion ticket-taking sale total. Finally, in 2018, it was announced that *Jurassic Park* was selected for preservation in the United

States National Film Registry by the Library of Congress for being "culturally, historically, or aesthetically significant."

And just look at the franchise *Jurassic Park* went on to spawn. Adjusted for inflation, Jurassic World is the seventh-highest-grossing movie franchise of all time (for more data see IMDB):

1. James Bond
2. Star Wars
3. Marvel Cinematic Universe
4. Harry Potter
5. The Lord of the Rings
6. Batman
7. Jurassic Park
8. Spider-Man
9. Pirates of the Caribbean
10. X-Men

Such statistics came with a cultural clout. Moviegoers the world over were seduced by the sight of the dinosaurs. Gone were the days of clunky stop-motion "T. Rex versus Trike" droll dinosaur battles to the "death." In its place were stunning technical achievements in visual effects and sound design. *Jurassic Park* is still considered a landmark in the development of computer-generated imagery and animatronic visual effects. I vividly recall seeing the film for the first time. Sitting in movie theaters, many of us were struck by the realization of what the ongoing CGI revolution might mean for presenting science on the silver screen. The dinosaurs were created with groundbreaking CGI by Industrial Light & Magic, the American motion picture visual effects company founded in May 1975 by George Lucas. The dinosaurs looked jaw-dropping.

As a result, we shared in Dr. Alan Grant's emotions as he first sees the brachiosaurus (I am happy to tell you it's pronounced brake-ee-o-saw-rus) move across the screen with such towering elegance (even if its haunting call is actually the sound of a whale and a donkey). Fictional paleontologist Grant seems so gobsmacked by the CGI, which finally does justice to the science, that he merely utters, "It's a dinosaur!" When John Hammond tells Grant that the Park also has a T. rex, Grant almost faints and barely recovers before we spy a herd of brachiosaurs and crested, duck-billed parasaurolophus at the lake. Later, Hammond declares of the Park's dinosaurs, "They'll capture the imagination of the entire planet." And so, they did.

A New Generation of Dinosaur Hunters

Without *Jurassic Park,* many dinosaur hunters wouldn't have a job. The movie led to an explosion in the public's interest in dinosaurs. And soon, by extension, it led to a boom in people wanting to research dinosaur science. How did this renaissance happen? The movie created a brand-new image of dinosaurs for the emergent generation. Dinosaurs were now dynamic, lively, and intelligent creatures. The explosion of public interest meant a lot more museums created dinosaur exhibits. And there was a huge increase in the number of colleges and universities running dinosaur-related courses. And that meant more money for the field of dinosaur science.

Many professional scientists believe that, were it not for *Jurassic Park,* dinosaur paleontology would still be a niche discipline. There would probably be just a handful of people studying it, and not a very diverse group of people at that. The *Jurassic Park* franchise revolutionized the field, and scientists reaped the rewards because there were so many people around the world who were galvanized by the movie.

As a result of this revolution in dinosaur science, one of the great ironies is that the *Jurassic Park* films soon became out of date. The task of this book is to compare the fiction of the *Jurassic* films with the facts we have learned about dinosaur science since 1993.

Jurassic Thought Experiment

Jurassic Park was an amazing thought experiment. Scientists also use thought experiments. German genius Albert Einstein was famous for them. In his youth, he mentally chased beams of light. The technique led to his famous Relativity Theories. *Jurassic Park* was a testing ground for discussing the scientific and ethical challenges of bringing back the dinosaurs. To create a theme park, and to discover how dinosaurs from different times would interact. Because T. rex didn't coexist with brachiosaurus. And neither of those two dinosaurs lived at the same time as Dilophosaurus or Velociraptor. Such creatures lived at different times, in different places, as we will discover later. But this is one of the things that makes *Jurassic Park* such a great thought experiment for science too. How *would* T. rex deal with the Velociraptor? How *would* these very different types of predators cope with a new and weird wilderness?

What's the biggest flaw in the way the *Jurassic Park* thought experiment was pictured? Lack of feathers. Forget, for now, the genetics of bringing dinosaurs back to life. The one big omission in the film's portrayal of its thought experiment is the absence of feathers on some of the dinosaurs, as we now know many dinosaurs were feathered. *Jurassic Park* did such a good job of picturing its CGI'd dinosaurs that most of the public would now probably find it weird to see these huge creatures feathered. Imagine being chased by a murder of feathered velociraptors. Because we know they had feathers. And we know they actually had wings. In my opinion, the fact that they actually looked like giant killer-birds makes them even more terrifying!

Dinosaurs Aren't Just for Kids

Who is this book for? You've probably noticed that there are so many dinosaur books for children that a library's worth of them would weigh the same as an Argentinosaurus. Sadly, there are comparatively few dinosaur books for adults. Especially popular science books aimed at the Jurassic franchise which, after all, is often the reason for a love of dinosaurs.

This book aims to right this wrong. We are, after all, living in a golden age of dinosaur science, a golden age of new dinosaur discoveries. It's time this tale was told against a backdrop of one of the most successful film franchises of all time. And, whether or not you want to relate to a childhood enthusiasm for dinosaurs, you will hopefully learn why dinosaurs are so fascinating, what makes them topical still, and why it's scientifically important to understand them.

So, read on, and experience the amazing story of the birth of the dinosaurs, how they evolved to world dominance, how some became gargantuan in size, how others grew wings and flew, and how the rest of them met an untimely end, at the peak of their dominance. It is a tale which paves the way for our modern world.

It's also a tale of some of the most amazing creatures ever to grace this tiny planet. Their global dominion over nature is a story that stretches over 150 million years of deep time, and the evolution of fantastic creatures that does justice to some of cinema's most incredible creations.

HOW WAS DEEP TIME DISCOVERED?

"Push up some mountains. Cut them down. Drown the land under the sea. Push up some more mountains. Cut them down. Push up a third set of mountains, and let the river cut through them . . . What is 40 million years? Enough time for a small predatory dinosaur to evolve into a bird. Enough time for a four-legged, deer-like mammal to evolve into a whale. And far more than enough time to turn an ape-like creature in eastern Africa into a big-brained biped who can marvel at such things . . . Such analogies hint at what deep time means—but they don't get us there. 'The human mind may not have evolved enough to be able to comprehend deep time,' John McPhee once observed, 'it may only be able to measure it.'"

—Keith Meldahl, *Rough-Hewn Land: A Geologic Journey from California to the Rocky Mountains* (2011)

In *Jurassic World: Fallen Kingdom*, much of the action takes place at Lockwood Manor. The Lockwood Estate, situated in Northern California, belongs to Sir Benjamin Lockwood, wealthy philanthropist and longtime friend and business partner of John Hammond. The grandest part of the Manor is its museum parlor, which is adorned with large dinosaur diorama displays, similar to Disneyland and real-world museums such as the Smithsonian National Museum of Natural History. Many moons ago, the dinosaur discoveries in the Smithsonian were housed in a Lockwood-like room known superbly as the Hall of Extinct Monsters. Today, that same space is known as the Deep Time Hall. And the scientific discovery of deep time, as we shall now see, is a tale of such drama it's well worth some kind of Steven Spielberg treatment.

The Great Chain of Being

The story starts in the medieval Age of Faith. This was a time when it was thought all things were created by some god, or gods, above. Just as a typical medieval town was walled-in, so was the medieval universe. A walled-in cosmos, bounded between Heaven and Earth, closed to the ravages of change and time. It was a fully connected cosmos, from the realm of God beyond the stars, through the nested planets carried along on spheres of crystalline perfection, and down to the dark and lowly corruptible Earth at the center.

This old universe was an ornate pageant of divine creation, known as the Great Chain of Being. Within the broad boundaries of this cosmos sat a cornucopia of creation: an infinite procession of links, stretching from God above down to the lowliest form of life. This scale of being, or *scala naturae*, was a strict hierarchical system, ranging from the highest perfection of the unchanging Spirit, who sat at the top of the chain, down to the fallibility of flesh at the core, mutable and corruptible "Man."

Every creature and object had a place in this great scheme of things. And each place was determined in a rather anthropocentric way, often according to its utility to humans. For example, wild beasts were superior to domestic ones, since they resisted training. Useful creatures, such as horses and dogs, were better than docile ones, such as sheep. Easily taught birds of prey were superior to lowlier birds, such as pigeons. Edible fish were higher up the totem pole than more dubious and inedible sea creatures.

Even aesthetics came into it. Beautiful creatures such as dragonflies and ladybirds were considered more worthy of God's glory than unpleasant insects such as flies and, no doubt, dung beetles. The poor snake languished at the very bottom of the animal segment, relegated as punishment for the serpent's actions in the

Garden of Eden. Some aspects of the Chain persist in popular culture today: the lion is still considered king of all existing creatures, the T. rex king of the extinct dinosaurs.

This explains a memorable scene in the movie *Jurassic World: Fallen Kingdom* where a lion faces off against a T. rex. This famous moment is symbolically saying that the T. rex is taking its place as the true king, roaring at the lion in a declaration of dominance. Within the movie, thanks to the science of genetics, the Great Chain of Being has been rewritten.

Limits in Space and Time

Historically, the Great Chain was, in part, a theory of biology, a theory of the generation of "sentient and vegetative creatures." As such, it was a hugely influential idea, until it was eclipsed by the theory of evolution many centuries later. But medieval notions of history left little room for the concept of evolution. The Christian sweep of time was a mere bite-size history, one that began in Genesis and ended in Revelation.

Nonetheless, Church scholars wanted to get a temporal grip on their history. How long ago had God created all this wonder? They began to add up the "begats," the long procession of scriptural births and deaths found in the Christian Bible. The fashion for doing so started with Eusebius, Chairman of the Council of Nicaea in 325 AD. Eusebius claimed that 3,184 years had elapsed between Adam and Abraham. Medieval German astronomer Johannes Kepler caught the dating bug and estimated the date of Creation at about 3993 BC. Even world-famous physicist Isaac Newton followed suit, putting the date at 3998 BC.

This rather dodgy method was raised to an "art form" by the seventeenth-century bishop of Armagh, James Ussher. He declared in 1658 that, "The beginning of time, according to our chronology . . . fell upon the entrance of night preceding the 23rd day of October, in the year of the Julian calendar, 710." That's

4004 BC to you and me. There is no doubt that some still believe such twaddle to be true. But the age-dating techniques of these Christian chronologists did have some lasting worth, it seems. They unconsciously paved the way for more scientific inquiries about the genuine extent of the past.

Learning to Read the Rocks

Science began to bury the Age of Faith. By the end of the 1700s, humans had made a start in securing their dominion over nature. Newton's system of the world rang out in the clanging new workshops of the world. The steam engine drove locomotives along their metal tracks. The first steamships crossed the great Atlantic. Transport magnates built bridges and roads. Telegraphs ticked intel from station to station. Cotton works glowed by gas. And a clamorous arc of iron foundries and coal mines powered this Industrial Revolution.

Speculation about planet Earth and its fossils grew steadily in the eighteenth century, along with a fascination for the natural world. And the new disciplines of geology and biology began to locate humanity's place in the depths of time, just as astronomy had charted our new position in cosmic space.

In developing industrial nations, such as Britain and Germany, engineers could open up the veins of the Earth deeper than ever before. Budding field geologists oversaw the excavation of strata, laid down over hundreds of millions of years of planetary history, as yet an untold story. The geologists soon learned how to read the rocks. As French naturalist George Louis Leclerc explained in 1778:

> Just as in civil history we consult warrants, study medallions, and decipher ancient inscriptions, in order to determine the epochs of the human revolutions and fix the dates of moral events, so in natural history one must dig through the archives of the world, extract ancient relics from the bowels

of the earth, [and] gather together their fragments . . . This is the only way of fixing certain points in the immensity of space, and of placing a number of milestones on the eternal path of time.

The Fossil Record

The geologists began to learn the language of the stones and started to decode nature's cipher. And with this understanding they amassed evidence of the long history of planet Earth. Take the English geologist William Smith, for example. Smith was consulting engineer for the Somersetshire Coal Canal. In 1793 he noted that the same strata were usually found in the same order and contained the same fossils. And so it was that the geologists realized the world's natural history could be gleaned from the fossil sequence hidden within the rocks.

But this soon spelled trouble for unsuspecting scientists. The fossil record started to churn out signatures of beasts no longer found on planet Earth. The evidence flew in the face of biblical accounts of natural history. What was worse, many of these beasts had no living counterparts. That was a problem for the faithful. They derived their Earth history from Genesis, along with the belief that all animate creation was born at the same time and that none such had become extinct.

Look Out for the Woolly Mammoth!

The Church fathers were far from convinced by the growing evidence in the fossil record. Give our lord God time, they suggested. Soon enough, His divine essence would ensure that living and breathing specimens of all assumed-dead species will materialize. Maybe in far-off lands to which the beasts had roamed in the years since the strata were formed. American President Thomas Jefferson was one such evangelist. Jefferson urged pioneers heading west to look out for the woolly mammoth. A pious but deluded naturalist even reported having heard one trumpeting through the dark forests of Virginia.

And yet the death roll of extinction grew. French zoologist George Cuvier helped found the science of paleontology. By 1801, Cuvier had identified twenty-three extinct species in the fossil record. The word "extinction" found a place in the lexicon of science, and also started to ring out in churches and chapels far and wide. Today we understand that 99 percent of all species that lived on planet Earth have since perished, including the dinosaurs. But that was a tale as yet uncovered.

The Terror of Time and Those Dreadful Hammers!

The new geology revolutionized the world. Its impact spread far beyond science itself, destroying established truths, and forcing everyone to confront the terrible extent of time. Famous philanthropist John Ruskin was moved to comment in 1851. "If only the geologists would let me alone, I could do very well, but those dreadful hammers! I hear the clink of them at the end of every cadence of the Bible verses."

Biblical literalists sought refuge in the Great Chain of Being. But the Chain was no more robust than its weakest link. In fact, the very wholeness of the Chain was taken as proof of the glory of God. There could be no "missing link," a term later appropriated by the evolutionists themselves. To understand just how fearful were the faithful of the idea of the "missing link," consider the words of English philosopher John Locke. Locke wrote in 1689:

> In all the visible corporeal world we see no chasms or gaps. All quite down from us the descent is by easy steps, and a continued series that in each remove differ very little one from the other. There are fishes that have wings and are not strangers to the airy region, and there are some birds that

are inhabitants of the water, whose blood is as cold as fishes
... When we consider the infinite power and wisdom of
the Maker, we have reason to think that it is suitable to the
magnificent harmony of the universe, and the great design
and infinite goodness of the architect, that the species of
creatures should also, by gentle degrees, ascend upward from
us toward his infinite perfection, as we see they gradually
descend from us downward.

The words of Locke clearly confirm the damage to be done by
any idea of extinction in the minds of the pious. As Royal Society
botanist Peter Collinson put it, "It is contrary to the common
course of providence to suffer any of His creatures to be annihilated." Similarly, the seventeenth-century English naturalist John
Ray feared that evidence of "the destruction of any one species"
would result in "a dismembering of the Universe . . . rendering it
imperfect."

The idea of extinction and "missing links" proved fatal to the
Age of Faith. Dinosaur remains were monstrous deviations, fell
portents of undesired change, and proof that something was very
wrong in a system supposedly framed by the hand of God. To make
matters worse, in December 1831, a young naturalist set sail on
The Beagle. Charles Darwin would soon become a chance locus
not only for the dissolution of the Chain, but also for a revolution
in science which struck at the heart of humanity itself.

Voyage of Discovery: Uniformitarianism and Catastrophism

Darwin was about to embark on a personal voyage with universal
themes. His journey would force the world to wake up to some
very alarming facts about the Earth's natural history. Now, a scientific theory needs a paradigm, a framework in which to develop
hypotheses and test them. Darwin's paradigm was an account of

evolution developed directly from the new geology. The foundation stone was gradual change: that the Earth is ancient and continues to change today, just as it did in the past.

At first, Darwin had little doubt about Scripture. Like many natural philosophers of the early nineteenth century, he considered all species to have been simultaneously and individually created. In short, as Darwin put it, he did "not then in the least doubt the strict and literal truth of every word in the Bible." But his observations on the voyage, and the lucid and vivid writings of the new geology, began to change his mind.

Like the geologists, Darwin believed that "The ruins of an older world are visible in the present structure of our planet." This "uniformitarian" belief held that all change, in geology and biology, was down to slow but unstoppable processes running throughout Earth's long history. In contrast, many Christian fundamentalists believed in "catastrophism." The catastrophist response to the evidence of the fossil record was the opposite of the uniformitarian. Where the uniformitarians like Darwin saw slow change as the way to explain natural phenomena, catastrophists saw sudden, near-supernatural upheavals, leveling mountains in a minute, heaving up seabeds at a sweep, dooming entire species in a matter of seconds.

Nonetheless, the Christian account of Earth history faced many problems on the biological front. The fossil record overflowed with that cornucopia of cadaverous creatures. Fossilized flowers that no one had ever seen in bloom, woolly mammoths and rhinos, and a whole host of bizarre beasts that looked like they had just walked out of Bosch's *The Garden of Earthly Delights*. And then there were the terrible lizards, of course.

The remains of many of these extinct animals were found in curious locations, places where they could not possibly have prospered: sea creatures on hilltops, polar bears on the equator, that kind of thing. The planet must have undergone massive upheavals, to say the least, for such profound changes to have happened in the short time imagined by the biblical literalists.

Journey's End

Darwin's *Beagle* became one of the most famous ships in history. And Darwin himself was confronted with a world rich in diversity. HMS *Beagle* sailed across the Atlantic, around the southern coasts of South America, returning via Tahiti and Australia, having circumnavigated the Earth. While the expedition was originally planned to last two years, it lasted almost five. The message Darwin brought back was clear. If the theories of geology and biology were right, then "the causes which produced the former revolutions of the globe" continued today. And that had profound consequences in time. The age of the Earth must be measured in millions, not thousands, of years.

The idea of deep time was born. Darwin delivered the true innovation of the age. He helped identify the evolutionary mechanism by which new species came to be. Darwin's magnum opus, *On the Origin of Species*, meant that the theory of evolution became the lifeblood of science. "He who ... does not admit how vast have been the past periods of time may at once close this volume," Darwin wrote.

For species to have evolved on Earth, the genuine extent of history was deep time, and not the six thousand years or so that was suggested in Darwin's day. Victorian biologists and geologists implied that the Earth was ancient. However, they did not prove it. That was done by physicists in the twentieth century. Their radiometric dating of the rocks—the technique for dating materials using naturally occurring radioactive isotopes—proved that planet Earth was billions of years old. Deep time was here to stay. Over to you, Mr. Spielberg.

WHY DIDN'T ALL DINOSAURS BECOME FOSSILS?

A crew huddles on the ground in the Montana Badlands, brushing and hand scraping at loose sand to slowly unveil a more than 70-million-year-old Velociraptor skeleton. Unfortunately for them, they don't yet realize that velociraptors are only found in East Asia, and it's their 110-million-year-old relatives, deinonychus, that are actually native to ancient Montana. Anyway, the skeleton is pretty much intact, with solid bones, a tail arched over its back, and its head bent backward with open jaw in a posture known as the "death pose." This whole scene is obviously how dinosaurs are found and dug up, right? Well, not exactly.

This particular vision of a dinosaur excavation in the first *Jurassic Park* movie has its downfalls, but the general agenda holds (i.e., it's possible to find the remains of dinosaurs by digging in the earth). However, considering dinosaurs existed all over the world's landmass for around 180 million years during the Mesozoic Period (252 million to 66 million years ago), shouldn't we literally be finding their fossilized remains everywhere?

Fantastic Fossils

The ancient remains of deceased life forms have long been found by people all across the planet. These remains included shells, plants, and animals, which were all mostly recognizable by their resemblance to living creatures, with the exception that they appeared to be made of stone. These petrified forms, extracted from the ground, were later given the name fossils, which comes from the Latin word for "something dug up."

Some of the most remarkable fossils were the huge stone bones from animals that were not entirely recognized as being extant on Earth today. These types of discoveries tempted the human imagination for millennia, prompting stories of dragons and other mythical creatures. Now, of course, we have a much better understanding of what they were and how they relate to the rest of life on Earth.

Among those early finds were also spiral forms called ammonites, or serpent stones in Medieval England, due to their appearance. These were quite common and are still found by amateur fossil hunters the world over. Ammonites are of immense importance as they act as index fossils, meaning their presence in the ground allows scientists to define and identify how long ago that particular layer of ground was formed.

Since those early discoveries, the study of fossils has changed a great deal, with new technology allowing the study of ancient life to be done in ways never before imagined. Each new fossil discovery provides added insight into the complex history and development of life on Earth, allowing us to develop a physical record of ancient life through the identification and classification of fossils. This physical record is known as the fossil record, and it has allowed humans to document how life has changed over time, with older fossilized life forms being replaced by newer ones that look more like modern life than the older ones.

The fossil record provides added support for the theory of evolution, presented by Charles Darwin and Alfred Russel Wallace in 1859. However, Darwin did remark in *On the Origin of Species* that there are gaps in the record where fossils of intermediate varieties of life had not yet been found. "But just in proportion as this process of extermination has acted on an enormous scale, so must the number of intermediate varieties, which have formerly existed, be truly enormous. Why then is not every geological formation and every stratum full of such intermediate links?" Basically, if new life forms are replacing (or exterminating) older life forms on an enormous scale, why don't we find a similarly enormous amount of fossilized remains littered everywhere throughout the ground? Well, a big part of it comes down to how fossils are formed.

The Fossil Factory

The study of how fossilization occurs is known as taphonomy. Fossilization can be achieved in a number of ways or preservation styles, depending on the environmental conditions in which the preservation occurred and on what the original specimen was made of. For instance, the majority of fossils tend to be comprised of harder remains like bones, teeth, and shells, as softer remains tend to decompose more readily.

Even though many fossilized dinosaur parts have been unearthed, 99 percent of fossil finds are actually from marine animals, where the remains can quickly become buried under layers of mud and silt that settle beneath the waters. Being buried has the advantage of stopping scavengers from devouring and scattering the organism's remains, while also preventing erosion and slowing the rate of decomposition by microorganisms.

Decomposition affects all dead life forms and is the result of chemical and physical processes instigated by oxygen, water, or living organisms (e.g., fungi, maggots, and microbes). In general, reducing oxygen, water, or temperature decreases the rate of decomposition by slowing down the activity of the decomposing organisms and reactions. This is why things preserve so well when frozen, when dried (like in deserts) or when buried (in oxygen-poor environments).

Some remains have been found of dinosaurs that died in dry places such as deserts, where they may have been quickly buried by the collapse of a sand dune, for example. This is believed to be what happened to a velociraptor and protoceratops discovered in 1971, who were buried mid combat. Nonetheless, the majority of dinosaur fossils are from animals that frequented or lived near rivers or lakes.

After an animal dies, its remains can end up underwater, whether through flooding or just dying in a wet location. With time, layers of sediment gradually build up above the creature's remains, squeezing them under the increasing weight. These layers of sediment get more compacted, driving out the water and eventually coalescing to become sedimentary rock, such as sandstone, limestone, coal, and shale. Throughout this millennia-long process, mineral-filled water seeps from the sediment and into porous gaps in the dinosaur remains. Depending on the type of sediment the remains become buried in, they will be saturated by different types of minerals. As the water exits the remains, it leaves behind these minerals, which accumulate and eventually solidify.

This process is called permineralization and it often involves the original bone material being left in place, with minerals filling in the porous gaps. There's also another process called replacement in which the mineral-filled water doesn't just fill in the gaps but also dissolves the solid material, such as bone, replacing it altogether and leaving a mineralized copy in its place. Permineralization and replacement are both types of petrification, which means "to turn into stone."

So, dinosaur fossils aren't actually bones but really distinctive formations of rock that took many thousands of years to materialize. They are most commonly found within sedimentary rock formations that were laid down while dinosaurs existed around 245 to 66 million years ago. Despite the many dinosaurs that lived in that time, the required circumstances for their fossilized preservation is very particular, which dramatically limits the chances of finding them. Nonetheless, according to paleontologist Steve Brusatte, we have found "Somewhere around 1,500 [species] so far. But probably there were millions that lived across the entire history of dinosaurs."

Dinosaur fossils aren't all composed of ancient bones, though.

Fossil Finds

There are different types of fossils accessible to investigators and they mainly fall into two categories: body fossils and trace fossils. Body fossils are the type that people usually envision when fossils are mentioned. They consist of actual plant or creature remains, such as the "Velociraptor" skeleton mentioned at the start of this chapter, or the famous mosquito found in amber. Fossils found in amber are limited to the very smallest organisms like insects, lizards, and scorpions so dinosaurs are generally an unlikely find. (Although, in 2016, a feathered tail of a small dinosaur was discovered in Myanmar, becoming the first dinosaur material to ever be found in amber.)

Dinosaurs leave behind more than just their bodies, though, and can also be recognized by the marks and trails they made while alive, such as where they crawled, walked, rested, fed, dwelled, or nested. These types of evidence are known as trace fossils, and the people who study these traces (whether recent or fossilized) are known as ichnologists. They even have a whole system of formally naming and classifying these trace fossils, called ichnotaxonomy.

The most publicly known trace fossils are probably footprints (called ichnites), whether alone or as a succession of footprints, called a trackway. These trace impressions are made on the surface of sediments but there's also subsurface traces such as burrows. Generally, these traces survive as petrified impressions in the ground, but in other cases they act like molds, which can fill in with sediment to create natural casts of the feature, known as cast fossils. Other interesting trace fossils are remnants of what creatures have used or produced in life, such as gastroliths (stomach stones), eggs, and even excreta.

Dinosaur species are mostly identified by body fossil remains, while trace fossils provide surrounding information about them. As trace fossils don't allow direct identification of a species, they can instead get classified as ichnogenera (trace fossil groups), based on features recognized from similar trace fossils. For example, a trace fossil may be associated with Hadrosaurs, but a more accurate identification might not be possible, so the Hadrosaur track gets categorized as an ichnogenus, such as Hadrosauropodus.

All of these discoveries contribute to the ever-growing fossil record, helping humans to develop our knowledge and understanding of ancient life on Earth.

Why Didn't All Dinosaurs Become Fossils?

When a dinosaur dies, there are a lot of things that need to happen for its remains to survive and be found. The animal's remains can become disarticulated and removed by scavengers, and water flow can also disarticulate rotting carcasses and transport the parts to different places, so creatures have to be buried relatively quickly after death to decrease the rate of damage and decay. This is more common in wet environments, so dinosaurs in those places are more likely to become fossils than desert dwelling creatures, except for on rare occasions.

In order for dinosaur remains to survive and be found, they also have to rest somewhere that can remain undisturbed over millions of years. Over time, land movements and erosion can expose parts of the fossils to the elements, leading to subsequent decay and damage from sunlight that can weaken and disintegrate bone. Forests don't make good places for revealing fossils either, as the roots and microbes there can hasten the breakdown and decomposition of the fossils.

The majority of creature remains would not survive all of these unique factors, or be accessible to humans at some point in the future. Those future humans have to also be looking within the right type of rock, typically sedimentary rock, which has to be within the geological layers that correspond to the age of dinosaurs (i.e., the Mesozoic).

Most dinosaurs died in far from ideal circumstances for preservation, so didn't become fossils, while the rest we just haven't found yet. However, as Darwin noted, there's an enormous number of creatures that have lived on Earth, so even an extremely low portion of that enormous number is still quite a few. Fortunately, the ability to interpret what little remains are left has become a science that has revolutionized our view of life on Earth and caused us to question the importance and fragility of our own existence on this planet.

HOW DOES AMBER HELP US PEER INTO THE PAST?

"Sometimes, after biting a dinosaur, the mosquito would land on the branch of a tree and get stuck in the sap. After a long time, the tree sap would get hard and become fossilized, just like a dinosaur bone, preserving the mosquito inside. This fossilized tree sap, which we call amber, waited for millions of years with the mosquito inside until Jurassic Park's scientists came along. Using sophisticated techniques, they extract the preserved blood from the mosquito and bingo! Dino DNA!"

—Mr. DNA, *Jurassic Park* (1993)

In the aftermath of a park gatekeeper getting mauled by a velociraptor, we meet an attorney arriving at the busy Mano de Dios Amber Mine in the Dominican Republic. He's speaking with the lead "digger," who takes him into the mine where we first catch a glimpse of one of the most amazing fossilized finds that make *Jurassic Park* possible. It's a one-hundred-million-year-old mosquito, entombed in a freshly mined chunk of orange-yellow amber.

There's no doubt that *Jurassic Park* raised the status of amber. Who would have thought that this fossilized resin that humans traded since the Stone Age would turn out to be the very thing to give the long-extinct dinosaurs a chance to roam the Earth once more? Well, that and a huge injection of money, science, and technical wizardry, which in the 1990s seemed tantalizingly close to what might actually be achievable in the real world.

When *Jurassic Park* was released, the value of insects preserved in amber spiked. It wasn't just people wanting peculiar jewelry and cane handles either. Paleontologists were also interested due to amber's excellent ability at preserving ancient life forms, along with new techniques for analyzing them. So, what's the big deal with fossilized amber?

Stuck in Amber

Despite being described as fossilized tree sap by Mr. DNA, amber actually starts out as a resin that oozed out of ancient trees. Sap can be clearer and waterier than resin and is produced by all trees as a part of transporting nutrients around the plant, whereas resin is thick and gloopy and mostly produced in conifer trees. This resin is produced in internal ducts or specialized surface glands in the plant and is thought to act as a defense mechanism against insects or fungi, particularly in response to damage.

When resin's chemically stable enough not to quickly degrade and be broken down by the weather, it can slowly start to solidify as some of its oils evaporate away. Any ancient animals that happened to be caught within this sticky resin will then become entombed forever. These prisoners of the amber are called inclusions. The surrounding resin then becomes fossilized in much the same way as other types of fossils, in that the resin becomes buried under wet sediments that are devoid of oxygen, protecting it from further degradation. With time and high temperature and pressure the buried resin undergoes chemical changes (polymerization) that eventually turn it into the fossil we recognize as amber. And while we imagine amber as being, well, amber-colored, it can actually come in a variety of colors including red, blue, green, and clear.

During the long process of fossilization, resin acts to dehydrate any trapped organisms. For many animals, this is only after their still-active gut bacteria and enzymes have started to rot them from the inside, leaving behind only the hard and now empty outer shell. Resin also has antibiotic qualities, so after it has soaked into the creature's tissues, it can provide a level of protection against fungus and rot, actively preserving the organism inside.

There are limits to what amber can preserve, though. For example, bigger, stronger animals (like the vast majority of dinosaurs) are much more able to free themselves from resin, which is why it's generally the smaller animals that are found trapped in amber, such as arachnids (e.g., spiders and scorpions), myriapods (e.g., centipedes), crustaceans (e.g., shrimp), and insects. Although, the biggest chunk of amber ever found measured 21.5" × 19.5" × 16.5", which would potentially be big enough to house a small cat, but most amber comes in chunks much smaller. Even then, amber still managed to preserve the tail of a small dinosaur from 99 million years ago, preserving bones, soft tissue, and exquisitely detailed feathers.

What can be found in amber is partly dependent on the size of the organism, but location is also important. Amber has been found all over the world with deposits in the Baltic region, Mexico, United States, Myanmar, Lebanon, Bavaria, Sumatra, and as seen in *Jurassic Park*, the Dominican Republic. Although, if you really wanted to find your own Jurassic mosquito, the Dominican Republic wouldn't be the way to go. Dominican amber is only 15 to 40 million years old, which is tens of millions of years after the dinosaurs became extinct. If you wanted to find 100-million-year-old amber, such as the type used in *Jurassic Park*, you'd have to visit places like New Jersey and Alberta in North America, as well as Myanmar (formerly Burma) and Lebanon. Amber fossils from these regions range from 90 to 120 million years old, meaning they originated in the Cretaceous Period when dinosaurs were still around.

Once we find a source of amber from a time in which the dinosaurs roamed, now we just need a mosquito, which (despite their annoying presence in life) are actually a relatively rare thing to find in amber. Being creatures that rely on water for reproduction, most are found in ancient lake sediments. In 2013, it was reported that the first-ever fossilized blood-engorged mosquito was found preserved in 46-million-year-old shale rock. Nonetheless, the oldest fossilized mosquitoes were actually found in amber, but only dating back to the Cretaceous Period, so no Jurassic mosquitoes as yet.

As a point of fact, the amber-imprisoned mosquito shown in the movie was a *Toxorhynchites (Tok-so-rin-ky-tees) rutilus*, also known as the elephant mosquito due to its long trunk-like proboscis. This is actually one of the only mosquito varieties that doesn't feed on blood so, sorry Mr. Hammond, that mosquito's a dinosaur DNA dud.

Why did Michael Crichton opt for a mosquito in amber anyway?

DNA and the Infamous Insect Inclusions

The oldest amber found dates back 320 million years, but the oldest animal inclusions are actually from about 230 million years ago, during the Triassic. This resin formed in trees growing in huge forests that were habitats for the many amber inclusions found today, consisting of animals, bacteria, plants, fungi, and chemicals that provide information about the surroundings at the time.

In the early 1980s, an entomologist (they study insects) named George Poinar Jr. was studying a 40-million-year-old gnat encased in amber. Working with his wife, they used an electron microscope to notice that the gnat's intracellular (inside the cells) structures had been preserved in the amber. It was further suggested by his colleague that DNA might also possibly be extracted from it. Almost a decade later, Poinar was contacted by Crichton, who was writing a new book and looking for a scientifically credible way of sourcing dinosaur DNA.

A few years later (a month before the movie's release!) the journal *Nature* published a paper describing how Poinar and his team had amplified and sequenced DNA from a 120- to 135-million-year-old weevil trapped in amber. To amplify the meager DNA sample, they used a technique called polymerase chain reaction (PCR), which can take a tiny section of DNA and make millions of copies so that it can be sequenced and analyzed. However, to get the DNA, they had to partially destroy the weevil, preventing any further study of features such as its morphology. Nonetheless, this was the type of scientific credibility that would help to underpin the Jurassic Park series, or so everybody thought.

In subsequent investigations by other scientists, where the sample and laboratory conditions were more carefully monitored, similar results were not obtained. This is an important aspect of science, where independent investigators verify claims by repeating a previous test or experiment. In this case, it turned out that the documented "ancient weevil DNA" was the result of cross con-tamination, mostly from a fungus, but also from a modern weevil that was also in the lab. The confusion occurred due to the PCR technique not being able to distinguish which DNA it amplifies, leading to a load of copies of the contaminant DNA as well. This contaminant DNA provided the false positive results.

Then, in 2012, any further hopes of finding dinosaur DNA in amber were dashed, when paleogeneticists (they study ancient preserved genetic material) studied the 600- to 8,000-year-old bones of extinct moa birds to work out the half-life of DNA. The half-life is how long it takes a given quantity to decrease to half of its value. For DNA, the quantity being measured are the number of bonds between the nucleotides making up the DNA's backbone. The researchers found the half-life of DNA to be 521 years and further calculated that even when stored at the ideal temperature of 23°F (-5°C), DNA would pretty much be completely degraded after about 6.8 million years.

The oldest genetic material ever sequenced was from a frozen horse that lived 700,000 years ago. Then, in 2019, a new tech-nique enabled scientists to obtain genetic information from the 1.7-million-year-old tooth enamel of a rhinoceros. Even though these discoveries are groundbreaking, none of them involved organisms preserved in amber, so perhaps any possible hopes for future dinosaur DNA may not be reliant on amber preservation.

Additionally, in 2017, researchers at Nagoya University in Japan studied how long it takes mosquitoes to digest human blood and render DNA unrecognizable. They found that after two days they could successfully amplify DNA fragments using PCR, but after three days "the human blood had been completely digested" with "no recognizable DNA fragments." If this holds for dinosaur DNA in ancient mosquitos, then they would have to be totally preserved within about two days of dining.

I put this to George Poinar who reiterated that, "The mosquito would have needed to fly into the resin just after, or soon after, it fed on the dinosaur. Just after feeding, mosquitoes go to a resting place to sit and remove the moisture from the blood. Trees are often chosen as resting places—fresh resin could be on the bark of such trees, making it possible for the mosquito to be embedded immediately after feeding."

Nonetheless, a mosquito's gut probably wouldn't be the best place to store DNA, as also commented by paleontologist Steven Brusatte. "DNA always breaks down really quickly, regardless of the scenario. And mosquito guts have active enzymes breaking down their food, so it is probably a particularly bad environment for DNA to be preserved!"

DNA from an ancient mosquito might never be a viable option, which certainly doesn't bode well for John Hammond's team of dinosaur de-extinction specialists. So, although it once seemed promising, it would seem unlikely that amber or mosquitoes can provide us with the holy grail of dinosaur DNA after all.

What Good Has Amber Done Us Then?

It's clear that amber has a unique ability to preserve features of ancient life, helping scientists to shed light on aspects of the past that were not accessible before. It preserves accurate morphologies of encased organisms, maintaining details down to the subcellular level. Unfortunately, amber can't protect DNA from the high level of degradation it would amass over the more than 66 million years since dinosaurs roamed.

Despite the letdown of amber-preserved DNA, amber has provided many gifts to the study of prehistoric life. This includes discoveries of lice-like parasites that infested dinosaur feathers, a 20-million-year-old flea with the plague, a 100-million-year-old insect taking care of its offspring, a tick with 220-million-year-old mammalian red blood cells in its belly, and, among other things, a harvestman with an erection.

A relatively recent rise in amber fossils from Myanmar has sparked off a spate of new and controversial discovery. As of 2019, more than 1,478 species had been discovered there, although it will take many years for researchers to work out how to capitalize on the growing collection of amber-entombed fossils fully and ethically. With more discoveries to come, amber will continue to provide more insight into the history of life on Earth.

DOES "LIFE, UH, FIND A WAY?"

Ian Malcolm: "John, the kind of control you're attempting is not possible. If there's one thing the history of evolution has taught us, it's that life will not be contained. Life breaks free. It expands to new territories and crashes through barriers. Painfully, maybe even dangerously, but . . . uh . . . well, there it is . . ."

Dr. Wu: "You're implying that a group composed entirely of female animals will breed?"

Ian Malcolm: "No, I'm simply saying that life—uh—finds a way."

—*Jurassic Park* (1993)

Life is extremely hardy, and its resilience is down to its remarkable diversity. A diversity that was built upon genetic information but fashioned by the ongoing pressures of the environment.

Life has had a long time to colonize this planet and it has done so quite effectively. It's been a long journey with many life forms falling by the wayside. And even though an estimated 99.9 percent of all species that ever existed are now extinct, a fortunate few survived to pass on their genetic baton and prolong their lineage in the relay marathon of life.

Despite the many hurdles, life has persisted, taking advantage of whatever chance benefits its various genetic codes have afforded it. So, how has life managed to find a way to survive and how does that relate to the dinosaurs of *Jurassic Park*?

Diversity: Life's Biggest Strength

Life's first amazing trick was coming into existence in the first place. Somewhere and somehow, on a planet composed of inanimate chemicals, life spontaneously arose more than 3.5 billion years ago. Scientists still aren't exactly sure how this happened, but it's generally believed that first there was an evolution of chemicals that led to the formation of various complex organic molecules. Interactions between these molecules eventually led to the emergence of the first cells with the biochemical qualities we regard as life.

A basic function of life is that it extends its presence by dividing and replicating its cells, whether for growth of an individual organism or for continuation of an organism's genetic material through procreation. This has given life the ability to resist containment in time as well as space and ultimately provided an opportunity for the pruning process of natural selection to take place. The resulting evolution of life forms, in line with the pressures imposed by changing environments, provided the diversity that made it possible for life to "break free" and "crash through barriers."

It hasn't been smooth sailing, though. Life has experienced a multitude of extinctions along the way, such as The Great Oxidation Event (GOE), 2–2.4 billion years ago, or the five mass extinctions that happened within the last 500 million years. (Humans have been considered as the cause of a potential sixth.) Nonetheless, in all cases life had developed sufficient diversity to survive these events and subsequently expand into new territories. Currently the lucky survivors of these catastrophes cover the entire globe, from more than 3 miles below the surface to 47 miles above it. (Yes, there is even bacterial life living in the stratosphere!)

Although, if ever there was an advertisement for life's ability to "expand to new territories," it would have to be the so-called extremophiles. These microscopic organisms represent some of the hardiest life forms that ever existed. They specialize in eking out an existence in places usually deemed inhospitable to life, where conditions are usually too salty, hot, cold, dry, acidic, alkaline, pressurized, or even radioactive.

Life is definitely adept at surviving, even if it didn't have a particular choice in which of its species would survive. For the individual life forms that are around today, it was mostly just dumb luck that they weren't wiped out. On the whole, though, life has persisted because it diverged into many different life forms that were as diverse as the habitats they would come to occupy. However, none of this would have been possible without life's ability to multiply through reproduction.

Cell Division: How Life Found a Way to Reproduce

At the core of all reproduction is the division of cells, whether looking at a bacterium, bird, or blue whale. All organisms are composed of one or more cells, which contain all the stuff an organism needs to survive in its environment. The first life forms were made of a single cell and were also prokaryotes (pro-carry-oats), meaning the cells didn't have a membrane-bound nucleus to house their genetic material. These prokaryotes, namely archaea and bacteria, comprised two of the three domains of all life and were the sole living inhabitants of Earth for more than a billion years. The extremophiles were mostly archaea.

To produce more of themselves, single-celled prokaryotes use a process called binary fission. This involves duplicating the cell's genetic material, and then dividing the cell into two new cells, both of which are identical copies of the original "parent" cell. Binary fission is a form of asexual reproduction, meaning it can happen without the fusion of two sets of genetic information. As binary fission is a relatively simple process, prokaryotes can multiply very quickly through it. For example, if a species can replicate every 20 minutes, then in just under seven hours one individual could spawn more than one million offspring. This is why bacteria, which are single-celled, spread so quickly.

Asexual reproduction also occurs in the more complex eukaryotes (you-carry-oats), which include all plants, fungi, and animals. Eukaryotes are organisms whose cells contain their genetic material inside a membrane-bound nucleus. They can be single- or multi-celled. When single-celled eukaryotes need to reproduce they use an asexual reproductive process called mitosis, which basically divides a cell, replicating its genetic material to produce two new identical copies. It's similar to binary fission except now there's also a nucleus to be divided.

When mitosis occurs in a multicellular eukaryote it still involves cell duplication but only of particular cells within the organism. This type of cell division is how all organisms grow and repair themselves. Cell division doesn't always lead to perfect replication of genetic information, though, and these imperfect copies are known as genetic mutations. When these mutations occur within cells that produce offspring, it can lead to differences between life forms and as such is one of the causes of life's diversity. However, the biggest instigator is the process of sexual reproduction. Brace yourself, we're going in!

The Origin of Sex

When it comes to reproducing a whole multicellular organism, as opposed to just single cells, life's preferred method is sexual reproduction. It's been estimated that over 99.99 percent of eukaryotes reproduce this way. Sexual reproduction is a process that relies on the union of genetic material (i.e., DNA) from specialized sex cells called gametes. This union of genetic material is known as fertilization and the resulting cell, called a zygote, contains a full set of chromosomes on which its genes are held. The zygote then divides and differentiates via mitosis to eventually produce all the various cells of an organism.

All cells of an organism (except the gametes) are known as somatic cells and these contain a full set of paired (referred to as diploid) chromosomes. Half of the pair typically comes from the mother and the other half comes from the father. Gametes are unique in that they only contain a single set of chromosomes, a situation known as haploid. The gametes are produced by a special form of cell division called meiosis, which divides a germ cell (a cell from which a gamete is derived) in half in such a way as to produce new cells containing a single set of unpaired chromosomes. During meiosis there is some shuffling of genes so that every gamete is genetically unique.

Different organisms have different numbers of chromosomes. Generally, each chromosome has a specific identity. For example, human gametes have 23 chromosome identities, one of which is the sex chromosome (i.e., the one that determines what sex an organism will be). The sex chromosomes in mammals and many other animals are designated as either an X or Y chromosome, while birds and some reptiles instead use a system of Z and W sex chromosomes. During fertilization in humans, the 23 chromosomes of one gamete pair up with the 23 from the other parent's gamete to form a complete set of 46 chromosomes in the resulting zygote.

In human females, every somatic (i.e., non-gamete) cell contains one pair of X chromosomes (XX), so during meiosis of the female germ cell, the resulting gametes (egg cells) each get one X chromosome. In human males, every somatic cell contains one X chromosome and one Y chromosome (XY), so during meiosis of the male germ cell, half the resulting gametes (sperm cells) have X chromosomes and the other half have Y chromosomes. Gametes and their associated sex chromosomes are the initial basis for the difference between the sexes and their union during fertilization is the main reason that life forms have sex.

So, when Dr. Henry Wu engineered the animals of Jurassic Park to all be females as a security precaution against unauthorized breeding, it would have applied a major handicap on life's natural ability to bounce back from adversity. Although, as it turns out in the movie, life manages to leapfrog the "unauthorized breeding" hurdle just fine.

Blame It on Kermit

The dinosaurs have escaped human confinement and there's no males to be seen. In the first instance, they probably wouldn't have been very concerned by this as their immediate problems would have been finding food, water, and shelter. Nonetheless, the very next morning, Alan Grant discovers a bunch of dinosaur eggs remarking that, "The dinosaurs are breeding," and offers the following explanation: "They mutated the dinosaur's genetic code and blended it with that of a frog's. Now, some West African frogs have been known to spontaneously change sex from male to female in a single sex environment. Malcolm was right! Look, life found a way!"

Discounting the problematic issues surrounding the blending of frog DNA with prehistoric dinosaur DNA, there are actually a number of extant species that can change sex. One example is the West African reed frog but on the whole this sex-changing ability is mostly observed in plants, gastropods (i.e., snails and slugs, etc.), and fish. For instance, clown fish can change from male to female when there is no longer a leading female in the group, while Wrasses can change from female to male in a process that takes anywhere from a few weeks to a few months, depending on the particular species.

Generally, these sex-changing animals are hermaphrodites, meaning they are born with both male and female reproductive organs and so have the ability to produce gametes (sex cells: i.e., sperm and eggs) of either sex. If both sets of organs function throughout its whole lifespan, an organism is regarded as a simultaneous hermaphrodite, meaning it exists as both sexes simultaneously. On the other hand, if an organism is born mainly as one sex but later changes sex as a natural sequence within its life cycle, it's known as a sequential hermaphrodite. This change can be triggered by internal or external factors such as age, size, or even the absence of a dominant breeder in their group, as with the clown fish.

In 2019, a study by US and Australian researchers found evidence for sex reversal in wild populations of green frogs (Rana clamitans). The frogs are regarded as intersex because they're observed as having the genetic makeup (genotype) of one sex but the physical makeup (phenotype) of the other. This indicates that they have changed from their original genetic sex and are now exhibiting characteristics of the opposite sex. Based on their observations the researchers tentatively suggest that sex reversal may be a relatively natural process in amphibians as opposed to just being the result of external factors such as temperature and natural or man-made chemicals in their environment. In this case (and for amphibians and reptiles in general), these sex changes only occur during the animal's larval stage.

Even though some frogs can change sex, if that trait were somehow encoded into dinosaur DNA, it would only convey the ability to change sex in dinosaur embryos (i.e., in the egg) rather than in older animals. So, maybe some other fluke of nature caused the reversal to kick in at a later stage of development, similar to a Wrasse fish. Well, for such a case to be viable, the dinosaurs would have to be hermaphrodites and there is no current indication of that possibility in the fossil record or judging by existing life forms. And even in the unlikely event that the dinosaurs were hermaphrodites, the transformation would still require weeks or months to occur rather than happening overnight like in the movie.

Instead of invoking sex change to introduce males, maybe we can find a reproductive process that doesn't require males in the first place.

Males Not Included

Males are often needed to provide sperm to fertilize eggs, although it is actually possible for eggs to be laid without fertilization. This is seen in some birds and reptiles, but the eggs generally fail to develop viable embryos due to having an incomplete set of genetic material. So, you might think that with no male dinosaurs on the island, there's no way for the dinosaurs to reproduce effectively. Well, it turns out that life in its wonderful diversity has already found a way around that, too.

For example, in hymenoptera (bees, ants, wasps), females can lay unfertilized eggs that develop into males with half a complete set of chromosomes (i.e., the males are haploid). This is a form of reproduction called parthenogenesis, which basically means virgin origin. This reproductive strategy is utilized by more than 2,000 species, about 70 of which are vertebrates. Depending on the particular mechanism, the resulting offspring may be haploid or diploid.

Parthenogenesis typically involves the female sex cell developing without the need for fertilization and is common in plants and invertebrates, while also being seen in fish, reptiles, and birds, although more rarely. For some snakes and lizards (such as cnemidophorus and teiidae) parthenogenesis is obligate, meaning it's the only option they have for reproduction. Although, there have been cases where creatures that usually reproduce sexually have reproduced parthenogenetically purely by chance. This was recently observed with Charlie the Komodo dragon at Chattanooga Zoo in Tennessee but has been witnessed before. It has also been seen among turkeys, although the embryos don't generally develop well, often not even making it to the hatching stage. Also, the turkeys that did survive had actually been selectively bred for increased incidence of parthenogenesis. Nonetheless, parthenogenesis can produce viable offspring.

In parthenogenesis, the sex of the offspring really depends on the sex determination system used by the creature. In the XX:XY sex determination system, the females (XX) lack the Y sex chromosome that's needed to produce a male, so parthenogenesis in those species can only lead to female offspring, which is the case with the female-only species, cnemidophorus and teiidae. However, for animals using the ZZ:ZW system, such as birds and Komodo dragons, the females (ZW) can lay eggs that develop into males (ZZ) or females (ZW) because the necessary copies of both chromosomes are contained within their DNA. However, in the rare cases when it's been observed, the offspring were always male.

It's believed that dinosaur chromosomes likely paired up in a similar way to birds, so if the dinosaurs did somehow reproduce parthenogenetically, it's possible that the offspring were only males. This would mean that those traces of little footprints leading away from the hatched dinosaur egg clutch observed by Grant in *Jurassic Park* all belonged to males, thus restoring the sex balance on the island, and alleviating the need for any subsequent parthenogenetic practices.

The Lysine Contingency

Assuming the dinosaurs managed to reproduce in the wild and maintain their longevity, there was another supposed problem they were up against. That is the so-called Lysine Contingency. As Ray Arnold from *Jurassic Park* explains, "The lysine contingency is intended to prevent the spread of the animals in case they ever got off the island. Dr. Wu inserted a gene that creates a single faulty enzyme in protein metabolism. The animals can't manufacture the amino acid lysine. Unless they're continually supplied with lysine by us, they'll slip into a coma and die."

Amino acids are the building blocks used by organisms to make proteins that perform a vast array of functions within the body, including growth and repair. Lysine is a genuine amino acid, one of the 20 used to make proteins, but in particular it is an essential amino acid, meaning that an organism's body can't produce it in sufficient quantities. This means that even without Dr. Wu's inserted gene, the dinosaurs still couldn't manufacture lysine. In fact, no animal can, which is why animals must consume it as part of their diet.

Lysine plays a role in growth and calcium absorption as well as in the production of collagen and carnitine, a compound that helps us to obtain energy from fat. When an animal suffers from lysine deficiency, symptoms can include nausea, fatigue, reduced growth, and reduced muscle mass. However, there is no evidence that they would slip into a coma and die.

Farmers often add lysine as a supplement in animal feed while wild animals obtain it naturally by eating the right foods, as explained by Dr. Sarah Harding in *The Lost World*. So, hundreds of millions of years before Ian Malcolm, before even the original dinosaurs appeared, life had already found a way around the problem of animals not synthesizing the necessary lysine.

How Does Life Find a Way?

Life has demonstrated that it can change sex, reproduce without males, and manage its own nutrient deficiencies alongside a world of other characteristics. Clearly, life *can* find a way, but it didn't come across these things overnight. It took billions of years of adaptations among countless generations of organisms. This provided the necessary diversity that allowed life to find multiple ways of surviving.

Things are different for individual species, though, where the options are typically limited to what the creature has in its own genetic arsenal. This limits life's ability to "find a way," as organisms can't spontaneously have a biological response that isn't coded in their DNA. Although the situation's a bit more complex in *Jurassic Park*, where the dinosaur's DNA has been blended with other creatures, leading to unexpected outcomes (i.e., sex change). This isn't a problem for life, though, as this is actually a part of how it became so diverse in the first place. Shuffling together genes through sexual reproduction, with the resulting offspring either dying, surviving, or thriving.

So, it's maybe not a case of life *finds* a way but more that life *has* a way, stashed within its diverse genetic tool kit. Within the tool kit lies solutions to things that have occurred in the past, might occur in the future, or might not ever occur, and whether they turn out to help or hinder a life form is just down to luck. Sure, having a huge number of test subjects and a great deal of time in which to diversify can help life as a whole, but for the limited number of dinosaurs in *Jurassic Park*, life would be extremely handicapped in its ability to find a way.

WAS DR. ALAN GRANT'S JOB A WALK IN THE PARK?

You know that point in a movie when the unsuspecting specialist gets the call to say they're needed for a top-secret project? It always made me think, *I wish I could be* that guy *and experience something unique and beyond the norm.* On deeper consideration though, the reality would seep back in. I just wasn't that special.

They're often chosen because they're at the top of their game and when the raptor poop hits the fan, it's their obsession with their specialism that can usually save the day. In *Jurassic Park*, Dr. Alan Grant is that guy, as are Dr. Ellie Sattler and Dr. Ian Malcolm. They each have their own specialism even though their survival seemed more to do with their personal attitude and luck than any particular job they had.

Nevertheless, Grant and crew were wandering through this park in the first place because they were regarded as some of the "top minds" in their fields. So, what exactly does Alan Grant do?

Paleontology

According to Dr. Alan Grant in *Jurassic Park III* (2001), "Dinosaurs lived sixty-five million years ago. What is left of them is fossilized in the rocks, and it is in the rock that real scientists make real discoveries!"

Paleontology quite literally means the study of ancient life. By ancient, this typically means older than about ten thousand years, which is so long ago that the ancient remains of life have long fossilized into rocky remnants. As such, paleontologists are largely concerned with understanding ancient life through the examination of these fossils.

By establishing the scientific study of fossils, paleontologists have provided insight into the ages of rock formations, mass extinctions, ancient animal behavior, past environments, ecosystems, climates, and major geologic events such as the movement of Earth's plates. Despite all this, it is still dinosaurs that have provided the greatest public appeal toward paleontology, capturing imaginations and enticing many a youngster to follow a career as a genuine dinosaur detective.

In the movie franchise, Dr. Alan Grant is a paleontologist who works at Montana State University and is credited with describing and naming the dinosaur "Maiasaura" with his colleague Jack Horner. In reality, Jack Horner is a prominent real-life paleontologist and one of the scientists that Grant's character is based on. Horner spent his time looking into the social behavior of dinosaurs while working at the Museum of The Rockies, which was affiliated with Montana State University. He and his colleague Robert "Bob" Makela were the real-life duo responsible for identifying and naming Maiasaura back in 1979.

Due to Horner's knowledge and expertise, he was one of the specialists called up by Spielberg to be a scientific advisor on his next "top-secret project," the *Jurassic Park* movie. So, Jack Horner was a real version of *that guy* we see in the movies (i.e., the person requested for their particular specialism). In a way, that means it's *his* job that I really envy when I get taken in by Grant's on-screen dinosaur park adventure. So, if I want to be *that guy* too, I just need to become a paleontologist. If only it were that simple.

Getting Qualified

To become a professional paleontologist requires years of dedication to learn the rudiments of the profession, which is mostly housed in geology and biology. We're talking a few years of undergraduate study, followed up by a higher degree, which is essential for those who want to teach. Then, for the major jobs in paleontology, a few years of research and a doctorate degree completes your round of formal education. After that, you're a bona fide specialist, ready to make your mark in the paleontology world as *that person* to go to for insight into your particular specialty.

Paleontology is a broad subject, covering about three and a half billion years of Earth history across multiple subdisciplines. For example, Dr. Ellie Sattler's specialism was paleobotany, meaning she studied ancient plants. There's also micropaleontology, which explores the remains of life forms so small that a microscope is needed to view them. Then, vertebrate and invertebrate paleontologists study the fossils of animals with and without backbones, respectively. If you wanted to be a dinosaur researcher, you would be looking at vertebrate paleontology.

However, you wouldn't be working in a silo, as it were. To produce a comprehensive picture of when, where, and how dinosaurs lived, you would need to collate insights from all of the many different areas of paleontology, as well as incorporating a wide variety of skills and practices, sometimes from seemingly unrelated fields of inquiry, such as computer modeling, statistics, and even engineering. So, what's the day job like?

Having a Field Day

The most recognized part of a paleontologist's work, and the activity that perhaps has the greatest appeal to the public, is fieldwork. It's an essential part of a paleontologist's training, taking up one or two months of the working year for some, with the main objective of gathering fossils for later study. That's what Grant and Sattler are up to with their crew on the dig site where they first meet John Hammond.

The cliché image of a wide-brimmed-hat-wearing field crew huddled over a partially exposed skeleton with brushes and a rock hammer is a sight commonly seen at dinosaur digs, to be fair. It's just practical when working in the field, although not everyone dresses that way. You're just as likely to find individuals wearing a cap, bandanna, hard hat, or no headwear at all. And it's not just qualified and practicing paleontologists who attend digs but also a cohort of students and often a few volunteers too.

A lot of work goes into planning and undertaking a scientific fossil hunting expedition. It's not a case of just stumbling across ancient bones and digging them out, although some fossils have been obtained in this way. This is real science, requiring measurements, observations, and reference to an evolving body of knowledge. As such, there are some common procedures to ensure that as much meaningful information can be extracted from the site as possible. Before any of this, though, you have to find the right location for the expedition.

For dinosaurs, you're mostly looking for sedimentary rock beds from the Mesozoic Era. These can be identified by using geologic maps. Next, you'll prospect the area, looking for fossils that have become partially exposed due to erosion or else are just lying on the ground. A popular place to look is at the bottom of a hill or cliff where eroded fossils may have dislodged and fallen down from a potential fossil bed above. You can't just start digging, though, you have to obtain permission from the landowner first. Once all the paperwork is in place, you can set up camp. This will be your home for the next few weeks.

You have to take all the tools and supplies you'll need, including tools to remove the surrounding rock. For substantial overlying rock (referred to as overburden) dynamite was used in the past, while jackhammers and rock saws are common nowadays. Working closer to the fossils, you might use rock hammers, dental picks, and brushes to reveal your precious finds. This can take weeks of grueling effort, sometimes requiring repeat visits in following years to clear a site more completely. You'll draw upon both your own and your team's knowledge and experience to classify the items where possible but ultimately, they will be bagged and tagged for transporting back to the lab or storeroom for deeper analysis.

The fragile fossils are often extracted with much of the surrounding rock matrix left in place. A glue is also used to help protect the crumbly fossil from falling apart when moved. The fossil is then carefully wrapped up in a protective jacket comprised of plaster over a layer of damp tissue, which shields the delicate bone from the plaster. The jacketed fossils can then be carried or wheeled out of the quarry if they're small enough, but for heavier items the team may have to drag them out on a sled or even in a helicopter. We get to see plaster jackets on-screen in *Jurassic Park III*, when Alan Grant leaves the tent to talk to Paul Kirby for the first time.

As a matter of good scientific practice, before fossils are removed from the earth, every single fossil's position and orientation are mapped and documented, along with information about the ground and surroundings. Any jacketed fossils are also given an identifying number so that they can be easily tabulated later by technicians who might spend many days or months cleaning the fossils, turning them into the marvelous, finished specimens we see in museums.

That's about 10 to 20 percent of the paleontologist's time accounted for, but what do they do with the rest of their time?

And the Rest of It . . .

After a grueling but satisfying few weeks in the field, the paleontologist returns home. But now starts the even more grueling work, which occurs behind the scenes, far removed from the field. This is how the other nine to ten months of their job is spent and where the major science of paleontology takes place—indoors. This can involve plying through existing research, investigating living animals and museum fossil collections, analyzing similarities between different life-forms, using mathematical modeling to describe and understand rates of evolution, and testing the age of particular rock layers using radiometric dating. The list goes on and with new advances more jobs appear on the list.

In the early days of dinosaur discovery, it was mostly wealthy, self-funded enthusiasts who indulged in paleontology or else people in the business of finding fossils to sell. The landscape is different now, as paleontology has become institutionalized within museums and universities. This is where the qualified paleontologist is most likely to find employment, where they might be teaching, doing research, curating collections, designing exhibits, preparing journal articles, presenting research at professional conferences, and communicating research to the public, including guest speaking on stage like Alan Grant in *Jurassic Park III*.

There are other options, too, like working with a geological survey or private industries such as those involved in petroleum or mineral and coal exploitation. And yes, some also work for companies that specialize in finding, restoring, and selling fossils to private collectors, although this is often less for scientific interest and more for profit. Although, for most dinosaur enthusiasts, the fortune isn't the pull: they discover dinosaurs for the advancement of knowledge. This doesn't mean money is not of concern, though. Far from it.

As any scientific researcher will attest, gaining funding is a huge and necessary precursor to undertaking research projects. It's mostly an administrative task involving a fairly detailed and lengthy application process to a government agency or private foundation. A 2011 article in *Scientific American* reported on a 2007 government study that found university faculty members spent 40 percent of their research time finding and making funding applications. To make matters worse, only about a fifth of those applications were likely to be successful. This is why in both *Jurassic Park* and *Jurassic Park III*, it's the promise of a generous grant that takes Grant away from his daily grind and off on a trip to those terrible islands off the coast of Costa Rica.

So, Was Dr. Alan Grant's Job a Walk in the Park?

People don't usually get to the top of their field by accident. It takes a huge amount of time and effort developing the skills and experience that enables them to be regarded among the best.

In the first place, Dr. Grant had to attend university and study all the way up to doctorate level where he earned a PhD. That's seven years at least. Then, consider the many more years spent doing research, fieldwork, and lecturing, along with all the funding applications, a good proportion of which wouldn't have even been successful. Seen in that light, Grant's work would've been less a walk in the park and more an arduous slog through a challenging but also satisfying landscape.

But what's past is prologue, and after all that hard work, he has now become *that guy*, embarking on an awe-inspiring job that literally involves him taking a walk in a park. It's fieldwork without the dust, long hours, and hard labor; all he has to do is take notes. However, as we all know too well, things promptly go downhill and what seemed like a cushy job turns out to be a harrowing ordeal that he wished to never repeat.

In the end, neither his job as a paleontologist nor his experiences on the Jurassic Islands were easy things to get through, so the answer is a firm nope. Dr. Alan Grant's job was definitely *not* a walk in the park.

HOW HAS RAPID PROTOTYPING TECH TRANSFORMED PALEONTOLOGY?

Billy: "You like computers, right?"

Dr. Grant: "I like the abacus, Billy."

Billy: "Meet the future of paleontology; it's a rapid prototyper. I enter in the scan data from the raptor skull, the computer breaks it down into thousands of slices, and this thing sculpts it. One layer at a time."

—*Jurassic Park III* (2001)

In this scene from *Jurassic Park III*, we are introduced to the rapid prototyper, a machine used by Billy Brennan to fabricate a 3D replica of a velociraptor resonating chamber. As the print finishes, he snaps it out and proceeds to blow through it, mimicking the call of a velociraptor.

Although a resonating chamber has never been obtained or even observed in velociraptors, something similar has been identified in some ankylosaurs and duck-billed hadrosaurs. In both cases, the relevant structures have been scanned and 3D printed and in one instance a sound artist even prototyped a full-size parasaurolophus resonating chamber that could be blown into by members of the public.

It's astonishing that we can print a physical copy of a structure that's totally concealed within a lump of rock. This game-changing ability is not just down to the 3D printing rapid prototyper though, there's also the technology that provides the scan data and the computers that process it. So, how have these digital technologies transformed paleontology?

3D Scanners

We live in a 3D world where things have real heights, widths, and depths. Whenever we take a picture with a camera, it converts that 3D world into a two-dimensional (2D) image that gets printed on a sheet or film or, more commonly nowadays, the image is digitized and displayed on a 2D screen. As a 2D image, the picture no longer contains real depth and all we're left with is our perception of depth based on visual cues in the image such as shadows or relative sizes and positions.

A rapid prototyper doesn't have our ability to perceive depth, so to reproduce a 3D image, it needs to be fed all of that three-dimensional information in the first place (i.e., height, width, and depth). This information is provided as data, captured via different techniques such as tomography (imaging by sections) or surface scanning.

As its title suggests, surface scanning technologies only give us 3D data about the surface of an object. Laser scanners are probably the most well-known example. These extremely accurate scanners can directly measure physical dimensions by reflecting a laser beam off an object and detecting the returning light with a sensor. They're ideal for objects larger than 10 millimeters, with resolutions down to about 50 micrometers. Laser scanners are also particularly good for targets that can't be moved from their location, such as trackways.

Another surface-based method used by paleontologists is stereophotogrammetry, which can analyze sets of 2D photographs to extract 3D information from them. The process relies on triangulation between different "lines of sight" that are calculated from particular visible points within the image. It's similar to how having two eyes helps us determine depth. It works on any sized object to whatever resolution the camera has, just as long as there are enough photos to cover the whole surface of the object. The fact that digital cameras are ubiquitous makes this technique widely accessible, extremely mobile, and relatively inexpensive.

Now, to image beneath the surface of an object requires another type of scanner, particularly one based on tomography. Examples include neutron tomography and optical tomography, but the most common technique used by paleontologists is the computerized axial tomography (CAT) scan, sometimes just called CT scans. This is likely how Billy Brennan obtained the scan data for his velociraptor resonating chamber.

CT scans use X-ray images to produce a high-resolution 3D image of an object, even if it's a fossil that's still covered in a rocky matrix. It works by taking multiple 2D image slices of the object from different angles. That's the tomography part. The data is then analyzed on a computer and used to construct a 3D representation of the object, which can be used to view its internal features. Using computerized tomography, paleontologists have revealed hidden structures within dinosaur skulls, such as their brain cavity, sinuses, blood vessels, and nerves. Studying structures like these can provide important clues about a dinosaur's potential physical and mental abilities.

Digital scanning technologies have transformed how data can be collected by paleontologists as well as expanding the range of things that can be imaged. However, the scans still need a bit of digital processing to neaten them up and make sense of them, which depends on a certain level of computing expertise. As such, paleontologists are becoming increasingly reliant on specific computing skills such as coding, as well as collaborating more with individuals proficient in digital technologies.

Computers

Computers are the ultimate digital technology and have absolutely revolutionized the study of ancient life, generating discoveries that could not have been made without them. One reason is their ability to crunch huge amounts of data and another is the versatility afforded by being able to tailor hardware and software to suit one's needs.

Computers allow scan data to be converted into formats that suit a desired output, such as a screen display or printer. For example, when using a CT scanner, the raw data is supplied to the computer as a set of 2D image slices called tomograms, whereas with laser scanners the data is supplied as information about thousands of points on an object's surface, known as a point cloud. This data is processed on the computer through the use of software packages that allow the scan data to be reconstructed as a virtual 3D model.

By using adequate computer packages, 3D models can be manipulated in several ways including viewing, editing, creation of simulations, and analyses. These interactive digital visualizations have proven to be absolute gold for paleontologists who are finding new ways to make use of these virtual models. As a result, computer-aided paleontology and virtual paleontology have become increasingly popular, as have the computer savvy individuals needed to run them.

For example, through the use of computer aided design (CAD) software, fossils can be digitally prepared. This can be done much faster than with actual fossil preparation, while presenting less risk of damage. Although, the process is largely dependent on the software's ability to distinguish the fossil from its surrounding rock matrix, which relies on special edge detection and image segmentation techniques. Similar processes aid in the differentiation of structures within the fossil (such as a resonating chamber), providing scientists with a way to virtually dissect fossils.

Virtual fossils are easier to work with and manipulate than the originals. For example, when arranging bones to explore a creature's possible anatomy and posture, there's no need to physically handle heavy or bulky casts or conversely tiny and delicate specimens that could be fiddly or break. Then, there's the unprecedented job of virtually correcting deformations that accrued while the fossil was buried. Also of value is the relative ease by which any missing parts can be restored. This can be done using digital scans from other specimens or by mirroring features from the opposite side of the body, just by making an on-screen selection and clicking a few buttons.

Investigators can even digitally add features like muscles and ligaments to construct a view of how a partial or whole ancient creature may have looked, functioned, and developed. These details can be tested and even tweaked to create hypothetical morphologies to test various ideas and scenarios. Additionally, by applying an engineering test called finite element analysis (FEA) to virtual bones or skeletons, it's possible to examine how the structure may have handled the various stresses and strains of everyday use. This includes analysis of bite force or the potential ability of bones or teeth to support different loads.

The use of digitized data is extremely useful, with an added benefit of allowing data to be quickly and easily shared with anyone in the world, such as researchers or for use in education or for someone who wants the data to create a 3D printed model for display or outreach. All that's required is a good computer processor and a bespoke software package.

3D Printing Technologies

Rapid prototypers are technologies that use digital data to create a physical 3D print of an object. This object may have been designed completely from scratch using CAD software, or it could be a direct (or modified) replica of a preexisting object (e.g., a fossilized dinosaur skull). As Billy briefly explained to Dr. Grant, the computer splits the 3D image into cross-sectional slices, which act like the blueprints for each layer that the printer will lay down. This data is then sent on to the rapid prototyper for printing.

There's a range of rapid prototypers currently in the market but the ones most used in paleontology use additive manufacturing (AM) processes in which 3D objects are built by adding on material layer by layer. Whenever someone mentions 3D printers, they're mostly referring to a machine that uses AM techniques to produce rapid prototypes.

The AM method most people are familiar with nowadays is called fused deposition modeling, also known as fused filament fabrication. This fast and relatively non-expensive process involves warming and then extruding a thermo plastic through a nozzle, which lays down the plastic layer by layer while tracing out the computed slice pattern. The plastic cools to form a solidified 3D object (i.e., the prototype). Additive manufacturing is well suited to production of one-off products, such as the scores of individual and unique vertebrae comprising a dinosaur tail.

The rapid prototyper we see in the movie works in a different way to AM; instead, it uses a technique called laminated object manufacturing, which was established a decade before the release of *Jurassic Park III*. This inexpensive method builds up laminate sheets of paper, plastic, or metal one layer at a time. Each successive layer is bonded onto the preceding layer and then cut to the desired shape with a laser or knife.

Another process available at the time was stereolithography (SLA) in which, one layer at a time, an ultraviolet laser is focused into a liquid consisting of light sensitive polymers. The light causes the polymers to react and form a solid, bound structure in whatever pattern has been traced by the focused laser beam. In 2000, a year before the release of *Jurassic Park III*, a study looked into the utility of stereolithography in paleontology, finding that the method faithfully replicated fossilized specimens, with the initial CT scan data being the most important variable affecting accuracy.

Overall, physical 3D printouts are useful for making replicas for museum displays or for use in education where valuable or rare specimens can be reproduced, providing wider access to students, researchers, or the public. It's also a bonus that they can easily be printed as rescaled versions of originals (e.g., large skeletons made smaller or tiny fossils made larger for easier handling and analysis).

Rapid prototyping supports a range of new experiments that were not possible before, such as printing out hypothetical body arrangements to test ideas about form and function, supporting a kind of experimental paleontology. This was done recently by researchers at Drexel University who built a robotic assembly of a dinosaur arm using 3D printed parts and physically simulated muscles and ligaments. The researchers hope to one day be making biomechanical models of whole dinosaur skeletons.

Maybe a form of Jurassic Park could become a reality, in the shape of mechanical dinosaurs modeled on real fossils and through the work of a roboticist rather than a geneticist.

The Future of Paleontology?

Billy Brennan wasn't lying when he suggested the rapid prototyper as being the future of paleontology. Although, to be fair, 3D printers are pretty much a sign of humankind's future in general. Technology is constantly evolving and it's taking paleontology along with it.

Where replicas and models were traditionally made by molding and casting or even sculpting objects by hand, now machines can be fed scan data from anywhere in the world and produce accurate replicas of any desired part of a fossil's exterior or interior. Indeed, it's been commented by paleontologist Scott Persons that digital dinosaurs are the future of paleontological collections.

Digitization helps in dissemination and education as does rapid prototyping, and with the range of available materials increasing and the cost of machines decreasing, it can only become more ubiquitous. In a world becoming more inundated with digital technologies and a growing global workforce becoming increasingly primed to use it, there's no doubt that the paleontologists of the future will find more ways to transform the field through new applications of existing and emerging 3D digital technologies.

HOW ARE DINOSAUR SKELETONS REPRODUCED IN MUSEUMS?

Do you remember that awesome feeling you got when you first saw a gigantic dinosaur skeleton? I don't mean the average-sized ones, but those huge behemoths like the alamosaurus and tyrannosaurus rex skeletons that get destroyed at the end of the first *Jurassic Park* movie. It had been awhile since I experienced that feeling of wonderment but a few years back I was reminded of it.

I was returning from Denver to the UK and had a changeover at Chicago O'Hare International Airport. Walking through Concourse B in Terminal 1, I came across the most amazing thing. It was a 40-foot-high replica of a brachiosaurus skeleton, just standing there, in the airport. Then, *pow*, in swept that feeling of awe again as I stood gaping up at it for a good five minutes or so. These displays never cease to amaze me.

I later found out that what I had seen was valued at $300,000 and obtained from Chicago's Field Museum in 1999. The museum was making space for the largest and most complete T. rex skeleton ever found. It was named Sue (after its discoverer Sue Hendrickson) and was bought for a whopping $8.3 million at a 1998 Sotheby's auction. Who'd have thought it; those bones don't come cheap! Sue is currently the world's most expensive dinosaur skeleton, but what goes into producing these spectacular dinosaur displays?

Getting Prepped

I used to think that dinosaur skeletons were bones that were dug out of the ground, carefully moved to a museum, then put in place exactly as they were found. That simplistic view turned out to be inaccurate in quite a few ways.

The way they're found isn't necessarily how they're displayed; the skeletons are actually rock and not bone and it's extremely rare that a complete dinosaur skeleton will be excavated due to the bones often being found disarticulated, broken, or with bits missing. Even the most complete T. rex ever found, specimen FMNH PR 2081, a.k.a. Sue, only contained 250 bones out of the roughly 380 bones it's believed that a T. rex should have.

After a fossil find has been extracted (which took six people 17 days in the case of Sue) and transported back for storage in a museum or university, for example, it must be removed from its protective plaster jacket and surrounding rock matrix in a process known as cleaning or preparing. The highly skilled and patient technicians who do this are called preparators.

Diligently peering through a microscope for precision, they carefully scrape and pick away the rocky matrix using a selection of instruments such as brushes, hardened needles, dental tools called air scribes (like mini jackhammers), mini sandblasters, and small grinding wheels. Less commonly, they may also use chemicals to help dissolve away some of the rock, but as with the tools, this must be done with extreme caution and attentiveness.

As the fossil is steadily revealed, the preparator makes use of carefully selected adhesives and consolidants (these are soluble glues that soak into the specimen to hold it together internally) to prevent the fossils from crumbling into pieces and also to repair any damage the specimen may have. This work is laborious and overall, the whole cleaning and preparation process can take absolutely ages. For instance, it took 12 museum preparators over 30,000 hours to prepare Sue's skeleton (and 20,000 hours to build the exhibit). Another record-breaking T. rex skeleton called "Scotty" (first uncovered in 1991) took decades to be prepared.

Once the process is done, molds of the specimen can be taken and the specimen given an official record number before being stored in the establishment's archives for future investigation. Alternatively, it may be used within a display. The molds, on the other hand, are used to cast copies of the original fossil. These casts are used in much the same way as the original pieces (i.e., stored for research or used to create displays on site or at other locations).

When the fossilized bones have been processed as required and there's not enough to make a full skeleton, what comes next?

Completing the Skeleton

There'll always be some missing information when reconstructing a dinosaur skeleton. In life, the dinosaur's bones are held in place by soft connective tissue such as ligaments, cartilage, and tendons. But, as these soft tissues decay away after death, the bones can become disarticulated and displaced. It is the job of the museum staff to reassemble the skeleton correctly, choosing a posture they think provides a good representation of how the articulated skeleton may have looked in life when all its connective tissues were in place.

To display the piecemeal fossil finds as part of a complete skeleton requires determining which fossilized puzzle pieces go where, then locating or fabricating any parts that are missing. This often incurs some educated guesswork but there are a number of ways it can be done.

They can model the missing parts based on casts from other finds of the same species, a method that can produce quite accurate results. They may also make internal and external scans of other similar fossil specimens, then 3D print adequately scaled replicas. This also makes it quicker and easier to obtain replacements if any pieces get damaged or worn over time. Another technique makes use of the fact that anatomical features are mostly symmetrical, so a missing left arm should be pretty much the mirror image of a present right arm. This generally works, although a 2015 discovery of an asymmetric styracosaurus skull has shown that there may be certain exceptions, as the horns observed on the left side of the skull were quite different from the horns on the right.

These methods all rely on having an existing fossil to extrapolate from but sometimes this isn't the case. If no existing specimens can be found to compare, the next best thing is to look at the anatomy of closely related species (e.g., other dinosaurs, birds, or reptiles). They may not be as accurate and so aren't as useful for scientific purposes, but for completeness of the public display, they'll usually do just fine. However, they have to be specially sculpted out of foam and clay, for example, to match the dimensions and characteristics of the display skeleton. Then, just like with fossilized bones, they can be molded and casted using materials like latex, resin, fiberglass, and plaster.

When all the parts are present, this prehistoric puzzle can be compiled by using the best current knowledge about the creature. But knowledge is a transient thing and rarely completely set in stone. Subsequent fossil discoveries and new analytical techniques can reveal more details about creatures that will inevitably lead to exhibits being updated accordingly. This recently happened in 2018 with the skeleton of Sue the T. rex. New research into T. rex provided fresh insights into the positions of some of its bones which prompted curators to adjust its displayed posture. As a result, its forelimbs now appear lower and closer together while its body appears bulkier.

These displays are often made from a mix of fossilized and replicated bones and updated in line with scientific knowledge, but how do they mount all those bones in the desired positions for display?

Mounting the Skeleton

The first mounted dinosaur skeleton was Hadrosaurus foulkii, originally displayed in 1868 at the Academy of Natural Sciences, Philadelphia. It was three stories high and built onto an iron framework. In the following years this new spectacle more than tripled the number of visitors to the establishment, spearheading the trend of mounting dinosaur skeletons for display. For the next 15 years it remained as the world's only mounted dinosaur display.

To ensure that displays are stored and erected safely, museums today often employ specialists called mount makers who design, make, and install the necessary supports known as mounts, brackets, or armatures. These supports provide the framework and fixings that hold the skeleton in the desired position and can include cables suspended from the ceiling and poles fixed into the floor or jutting out from a wall. The particular arrangement depends on the weight of the skeleton, which is determined by how much fossilized parts it contains.

If you look at Sue's skeleton, you'd be hard-pressed to tell the fossilized parts from the replicas. But even though they all look pretty similar, the fundamental difference is their weight, with the rocky fossilized parts being substantially heavier. This is true of Sue's skull, which weighs about 600 pounds. At such a weight, it would be quite difficult and dangerous to support it up high on the end of an iron or steel framework, which is why the museum opted to use a lighter replica skull for the skeletal display, with the heavy fossilized skull placed in a separate display case.

The main objective of the mount maker is to enhance the safety and viewing pleasure of the public. However, they also have to protect the displayed specimen from external damage as well as support it in a way that doesn't exacerbate any of its own inherent structural weaknesses. This is especially important when genuine fossils are being used. And as every part of the dinosaur skeleton has unique dimensions, mounts must be specifically fabricated to fit them, as well as orient the part correctly within the particular space.

To achieve this, they individually secure the skeletal elements onto a supporting frame that defines the desired posture of the animal. Each skeletal element might be connected using grub screws or by looping parts of the frame around them to make a cradle, similar to how a precious stone is held within a ring. Incomplete fossil bones may even be directly pinned to replica parts to make them appear whole within the exhibit. And to make fixings as discrete as possible, they may be fabricated to fix onto the object in a way that conceals it from view, as well as being recolored to match the exhibit if needed.

The materials used must also be considered carefully. Not only must they be strong, but they also have to be malleable enough to be worked into the desired shape. Typical materials include brass, aluminum, some steels, and acrylic plastic. They might even use synthetic casting resins that can function to support uneven objects and distribute contact loads evenly.

Mount making is a serious business and these guys don't take their jobs lightly. They bring together a range of different skills and disciplines to ensure that exhibits are safely and professionally mounted, whether at a museum, science center, airport concourse, Jurassic Park Visitor Center, or private collection such as Lockwood Manor's.

The Not-So-Final Resting Place

For more than 150 years, people have been mounting dinosaur skeletons for display. Skeletons that were discovered, dug up, and diligently prepared for subsequent analysis and potential mounting.

In most situations, only partial skeletons were found. While displaying a piecemeal skeleton may be scientifically accurate and representative of a find, it doesn't carry the same appeal to the public, so museums fabricate suitable replicas of the bones to bolster their displays, with some parts being genuine fossils and others being replicas. These are all mounted together on a specially made frame by mount makers who can install the skeleton safely as part of an engaging and emotionally rewarding exhibit. So, the next time you come across a dinosaur skeleton, maybe take a minute or two to look behind those bones to see some of the work these guys have put in.

There's a great deal of background work that contributes to the awe-inspiring reproductions of ancient dinosaur skeletons. Maybe that's what adds to their high monetary, scientific, and emotional value. It may not all be real dinosaur bones on display, but who cares, for the wonderment is the same. And for the few dinosaurs whose remains are displayed, it's almost unfathomable how more than 66 million years after their death, it would turn out that burial wasn't their final resting place after all.

HOW MANY DINOSAUR SPECIES ARE THERE?

"Dinosaurs and man, two species separated by sixty-five million years of evolution have just been suddenly thrown back into the mix together. How can we possibly have the slightest idea what to expect?"

—Dr. Alan Grant, *Jurassic Park* (1993)

"There's over 1,500 species that have been found and people are finding new ones all the time. It's really an incredible moment that we're all in right now because somewhere around the world right now somebody's finding a new species of dinosaur on average about once a week so we're getting about 50 new species a year. This is the Golden Age and they're coming from everywhere."

—Steve Brusatte, BBC Radio 4's
"In Our Time: Feather Dinosaurs" (2017)

Still Counting

The 2018 video game *Jurassic World Evolution* is based on the movie *Jurassic World*. The main focus of the game is the dinosaurs. Having lived for hundreds of millions of years during the Mesozoic Era, dinosaurs are an amazingly diverse group ranging from modern birds to species such as tyrannosaurus, brachiosaurus, and velociraptor, all brought back to life by InGen and the Hammond Foundation. During the game, the players are given the task of populating their dinosaur park, making sure of the park's success, and protecting the park's visitors in the process. *Jurassic World Evolution* was initially launched with just thirty-seven unique dinosaur species, with another five species added in subsequent downloadable content. But how many dinosaur species did nature herself conjure back in the old Mesozoic Era?

We speak elsewhere in this book about the fact that dinosaur science began in 1842. That was the year when British anatomist Richard Owen first coined the word "dinosauria." And we also mention that in 1887, only years before Owen's death, the very idea of dinosaurs was called into question. Well, how times have changed. For now, in the early twenty-first century, we're living through a golden age of dinosaur paleontology. Doctor Alan Grant would have a field day.

Fossil hunters are finding more dinosaur bones now than ever before. In fact, most weeks, someone somewhere about the globe is finding new species. So, the current rate of dinosaur discovery is around fifty species a year. And, as unbelievable as it may sound, it's been at that same rate for the last decade or so.

Why the sudden dinosaur deluge? Well, for one thing, there are so many more paleontologists around the world looking for dinosaurs these days. In lands like Brazil and Argentina and China, young scientists are being trained in their national universities and museums to hunt down dinosaur bones. And today's hunters are a far more diverse group than in the days of Richard Owen. Now there are many young women in the field, making today a far more exciting time for discovery.

Technology

Technology has also made a huge difference to the dinosaur count. In the last two decades, technological progress has been great. It's now normal for scientists to CAT (computerized axial tomography) scan dinosaur fossils. The CAT scan allows paleontologists to see inside the heads of dinosaurs. To see what their brains may have been like. To assess the nature of dinosaurs' sensory organs. And to gauge how smart dinosaurs actually were. As we describe elsewhere in the book, scientists also use computer animation software to gauge aspects of dinosaur locomotion to work out how fast they ran, how high they held their necks, how they fed, and the ferocity of their bite. All this work helps to uncover the real science behind the *Jurassic Park* and *Jurassic World* movies, and also helps determine one species from another when a new discovery is made.

One aspect of dinosaur science that has really taken off is the question of the rise of the dinosaurs. Huge advances have been made in this area. As we detail later in this book, we've known about the demise of dinosaurs for decades. Going back to 1980, in fact, when father and son scientists Luis and Walter Alvarez first suggested the idea that a rogue comet had caused the dinosaur catastrophe. And yet the topic of the rise of the dinosaurs has only been known in detail for the last 10 to 15 years.

This revolution in knowledge has occurred because so many new fossils have been found. These findings, of Triassic-age dinosaurs from so many regions of the globe, have uncovered an unexpected story. Dinosaurs didn't just stomp and spread about the globe from the very start, like some pandemic. (Not that pandemics stomp, of course.) They weren't any better than the other beasts alongside them in the early days after the Permian extinction on Pangea. Rather, the emergence of the dinosaurs was a protracted rise to dominance, which took around 50 million years.

Getting Down to Dinosaur Details: Welcome to the Mesozoic Park!

How did the dinosaurs become so diverse? Working out what counts as a distinct species is no easy matter. Scientists are argumentative by definition. And getting paleontologists to agree on a definitive list of dinosaur species is tricky. But let's first look at how many dinosaur species we think there were. Some studies try to work out the total diversity by using something called the species-area effect. In other words, if we know how many species there are in one particular region of the world, then we can extrapolate how many may have existed worldwide. This method suggests that there were up to one thousand dinosaur species at the end of the Mesozoic, 66 million years ago.

One thousand dinosaur species seems to be a reasonable esti-mate. Consider counting all of the Earth's living land mammals weighing more than 2 pounds, which is the size of the smallest dinosaur. Then, add in all the extinct species from the past 50,000 years or so. This would include woolly mammoths, giant kangaroos, ground sloths, and the like. You'd come to a similar total of one thousand species.

But this is merely the number of species extant at one point in time. And the dinosaurs were around for huge swaths of deep time, as we know. Over the course of their reign, dinosaur species were continuously evolving and going extinct. That's what evolution *is*, and what it *does*. So, bearing this in mind, let's come up with an estimate for those swaths of deep time. If we assume that one thousand species of dinosaurs lived at any one time, and that the species turned over every million years or so, we get a grand total of 160,000 species. That's a lot of dinosaurs.

And yet it is, naturally, a very rough guess. The 160,000 dinosaur species estimate depends on lots of assumptions. For example, just how many different species can the planet support? Another is, how quickly did dinosaurs evolve and disappear to extinction? Looking at such assumptions again, imagine we assume a diversity of only 500 species and a more gradual dinosaur turnover—species lasting 2 million years, for instance. In this case, we get about 50,000 species. At the other end of the estimates, with a diversity of 2,000 species in the lush, warm climate of the Mesozoic, and assuming the species lasted a mere half a million years, we get an estimate of over 500,000 species. What a real-life Mesozoic Park *that* would be! In short, however, we can safely say that there were between 50,000 and 500,000 species of dinosaurs.

Why So Many Species?

Dinosaur dominance can be explained by three main scientific ideas of biological evolution: specialization, localization, and speciation. For one thing, dinosaurs were specialists. This means that they were able to specialize by taking advantage of nature's various niches. And that meant different dinosaur species could coexist without competing. For example, in western North America, T. rex existed alongside the smaller meat-eating dromaeosaurs. And gargantuan long-necked sauropods ate alongside horned ceratopsians, which grazed on flowers and ferns. There were smaller plant-eaters still, the likes of ornithomimids and pachycephalosaurs, alongside heron-like fish eaters, and anteater-like insectivores.

There was even specialization *within* a niche. For instance, as T. rex was large with humungous jaws and stocky limbs, it was perfectly suited to preying on the almost stop-motion movement of the ambling but heavily armored triceratops. Whereas T. rex's relative, nano Tyrannus, was somewhat smaller but had the added advantage of the long legs of a Kenyan marathon runner (not literally), with which it was able to chase down faster prey. Such specialization meant that, according to recent studies of the fauna, up to 25 dinosaurs could live alongside one another in a single habitat.

Now, consider the idea of localization, the fact that different regions and habitats had different dinosaur species. For example, Mongolia had a particular set of creatures, Tyrannosaurs, ostrich dinosaurs, and duckbills, all inhabiting a lush delta which flowed through a prehistoric desert. Meanwhile, merely a few miles away, small-horned dinosaurs and parrot-headed oviraptors colonized the dune fields.

This localization pattern was also true across continents. Different dinosaur species inhabited different parts of North America, for instance. In fact, between continents, the differences are far more marked. Just one example among many: in the Late Cretaceous, Asia and North America were lands dominated by Tyrannosaurs, horned dinosaurs, and duck-billed dinosaurs. South America and Africa, however, lands cut off by oceans for millions of years, had an entirely different set of dinosaur species. Here, rather than Tyrannosaurs, the horned abelisaurs were peak predators. And, rather than duckbills, the long-necked titanosaurs were the peak plant-eaters.

Speciation: Thousands Left to Find

Finally, there's speciation. Dinosaurs evolved new species with incredible speed. Scientists have used radioactive means to date the rocks that contain dinosaur fossils. From this data, they can calculate how long dinosaur species lasted. Take, for example, the rocks of the Hell Creek Formation in Montana. They were laid down over a period of around two million years. At the base of these rock strata, paleontologists find the species Triceratops Horridus (horridus is Latin for "rough"). But, at the top of the strata they find a second species, Triceratops Prorsus (prorsus means essentially "pointing forward") which evolved from the first triceratops species.

This evidence suggests that a dinosaur species lasts up to a million years. That's rather a short time, geologically speaking. Indeed, similar studies of other rock formations, also of horned dinosaurs, suggest that other species were also relatively short-lived. For example, down in the badlands of Dinosaur Park, Canada, can be found fossils that suggest three different species of dinosaur. The first is replaced by the second, and the second by the third. And all evolved within two million years. So, dinosaurs evolved relatively rapidly. No doubt driven by shifts in our planet's oceans, climate, and continents. And, naturally, the evolution of *other* dinosaurs was an influence. In short, the story for the dinosaurs was evolve . . . or die.

We may never know precisely how many dinosaurs roamed the globe in the Mesozoic. It is, of course, rare for a creature to conveniently collapse and die, and to fossilize to be preserved. Given such rarity, you can appreciate that tens of thousands of species, if not hundreds of thousands, are likely lost to science forever. This makes it even more amazing that the rate of dinosaur discovery has increased over recent years. The great majority of the dinosaur species that have ever lived are lost forever. And yet there are still thousands left to find.

HOW DID THE DINOSAURS (MOSTLY) DIE OUT?

"When an impact event happens, the enormous kinetic energy of the [comet]—a massive body traveling at high speed—is partly transferred to the Earth, but largely converted into heat and sound, creating pressure waves traveling radially outward from its center, similar to that of an atom bomb. As well as this blast created on impact, large amounts of dust can be kicked up into the atmosphere—like a nuclear winter—blocking out sunlight and preventing plants from photosynthesizing, which has a knock-on effect to much of the rest of the ecosystem."

—The Institute of Physics's resource on cometary impacts (2015)

Dinosaur Chase

[*Jurassic World: Fallen Kingdom*: Isla Nublar, a satellite island off Costa Rica in the North Pacific Ocean]

The once-sleepy volcano erupts. Dark and dense smoke billows into the tropical island sky. Fireballs of rocky lava are catapulted into a nearby canopy of trees. Suddenly, we hear Owen somewhere in the distance yelling "RUN!" Then, we see him across a familiar field. Emerging from the volcanic smoke, he's barely one stride ahead of a cornucopia of dinosaur creatures that hurtle headlong toward the camera. It's mayhem—like an industrial mega-mix of "the animals went in two by two," but with explosive special effects.

Owen, Claire, and Franklin tear through the trees and the island flora to keep one step ahead of the stampede. Just behind our group, an ankylosaurus and a brace of brachiosaurs come crashing through the canopy, as our focus shifts from humans to brachiosaurs to volcano. To avoid being trampled underfoot, our trio take refuge next to a huge fallen tree and an abandoned gyrosphere. We get a better look at the majestic chaos, as a bull-horned carnotaurus roars against the backdrop of an even bigger eruption.

The mad parade continues. Owen makes like Bolt, as Claire and Franklin, in the rejuvenated gyrosphere, race alongside the charging creatures as terrified pterosaurs swoop overhead. Moments before the entire scene is swallowed whole by the black pyroclastic flow of the smoke, the gyrosphere plummets off the edge of a cliff and into the tropical sea. As we bear witness to the pathetic sight of the dinosaurs choosing death by water over death by fire, Owen manages to help rescue Claire and Franklin from a similar ruin.

Later, having escaped to the final boat to leave the fallen kingdom, Claire and Owen look back to the dock, as the lava and pyroclastic flow barrel down over what's left of the island. From this vantage point, we share in their last parting and poignant shot of Isla Nublar. Just visible through the smoke, we see a scared and desperate long-necked brachiosaurus left stranded on the dock, bellowing after the departing boat, unable to escape the volcanic armageddon. It raises up onto its hind legs before it finally fades into the glowing smoke. Soon, Isla Nublar will be lifeless.

Worse is to come. For that's not just *any* old brachiosaurus. That's the same gentle giant that greeted Alan Grant on his first trip to Jurassic Park. That's the awe-inspiring and majestic creature we spied as John Williams's score swelled, and cinema was testament to the wonders that science and nature can conjure. In short, the volcano scene in *Jurassic World: Fallen Kingdom* becomes a re-run of the actual demise of *all* dinosaurs. As Owen and Claire watch, the mass extinction becomes personal in the form of the sole brachiosaurus, and we wonder whether they may be projecting emotion onto a beast that is basically inhuman. And yet, in some senses, the brachiosaurus *becomes* human through projection.

After all, we may one day share its fate. But compared to the actual demise of the dinosaurs, the fictional drama on Isla Nublar is next to nothing.

Extinction Day

Planet Earth's deep time has seen five mass extinctions, those widespread and rapid decreases in the biodiversity of our world. Sixty-six million years ago, one of those worst days in history ended the 150-million-year rise of the dinosaurs. The tyrant lizard king, tyrannosaurus rex, would have been there to witness the devastation.

The day would have dawned like any other in the Cretaceous. Picture the scene. Huge swathes of conifers stretch to a distant green horizon, punctuated in part by the subtlest color from magnolias and palm flowers. The soundscape is equally evocative. We just about hear the distant roar of a surging river on its way to the western coast of North America beneath the din of a hundreds-strong herd of triceratops.

A pack (or perhaps a "terror!") of T. rex might well have been on the hunt that day. As they stalked in silence through the dappled light of the forest, we see the shifting shadows of smaller creatures who give ground to the tyrant of the food chain. Like the stampeding cornucopia in *Fallen Kingdom*, we are treated to our own host of dinosaur species: tank-like ankylosaurs and dome-crowned pachycephalosaurs rummaging through the trees, droves of duckbills munching on flowering flora, and raptors bedevil smaller mammals and lizards across the forest floor. Overhead, the silhouetted forms of species make their way across the Cretaceoun sky. Some feathered creatures flap their wings while others are lucky enough to glide on a zephyr in the hot and humid morning. The pipes and purls, the sonorous booms, lows, and ultimate roars make the most wonderful chorus of the Cretaceous.

Dinosaurs Don't Stargaze

Then, the day takes a turn for the sheer diabolical. The keenest of the creatures, perhaps the smartest of the rexes or the raptors, may have noticed a change in the Cretaceous heavens of late. For the last few weeks, there has been a new light in the sky. Another glowing disc, much fainter than the Sun, but gathering size and brightness over the days. Not that the creature could make much of the cosmogony of an approaching comet, of course. If comet it was, rather than asteroid. Besides, this "comet" had a habit of disappearing for hours on end, and the creature would go back to her usual business. Stargazing is not known to be a major concern of dinosaurs.

And yet, on this particular fateful day, should that curious creature glance upward, it might have spied something quite dramatic. Not only was the bright disc back in the sky, but it was now quite huge, its luminous glow dominating the southeastern sky.

Suddenly, a blaze of pure light. No sound. Just a fleeting flash that lights up the sky. A keen observer would see that the disc has now gone, and the sky has become a kind of cornflower blue. Then, another blaze of light, far more ferocious than the first. So ferocious, in fact, that the blaze is bright enough to burn into the retina. And yet no sound accompanies this spectacle. No beating of wings, no bellowing of brachiosaurs. The ancient world seems to hold its collective breath.

Prehistoric Rock and Roll

Calm becomes storm. Within seconds the ground beneath the dinosaurs begins to shake, thunder, and roll. Solid earth begins to flow like the surface of the sea. Waves of energy run through rock and soil, resulting in the same rise and fall as the ocean. It's as if a leviathan lurks beneath the very crust of the Earth. Reality seems to be on repeat. Anything not rooted in the ground continually rises and falls, as if the planet's crust has become elastic. And so, the smaller creatures, mammals, and lizards as well as dinosaurs are thrown into the sky only to come crashing down wherever they fall.

Nor do the larger dinosaur titans escape this fate. Huge tyrannosaurs and triceratops are also heaved into the air. Moments before they had been rivals at the top of an ancient food chain. The arrival of the comet has reduced them to bit players in a drama of planetary life and death. And so, Tyrannosaurus rex, the colossus of this continent, becomes a mere effigy in the ensuing storm. The forces of the cometary impact are capable of snapping legs, breaking necks, and crushing the skulls of even the peak predator.

The devastation wrought upon their world looks like the aftermath of war. Those that survived the initial impact make their way among the dead, as the sky turns once more. Blue becomes orange, orange becomes red, and the red gets angrier and brighter. It's as if the Sun has suddenly gone Red Giant and the light of the world has switched to scarlet. It's not long before all of reality looks rufescent.

A Hard Rain's Gonna Fall

That's when the rains begin. A rain not of water but of fire. Like the lava-bombs launched by the fictional volcano on Isla Nublar, scalding smithereens of glass and rock rain down on the surviving dinosaurs. The smithereens that fall are bean-sized beads that sear into the burning skin of the dinosaurs below. But the smithereens don't just claim their own victims to add to the day's dinosaur body count. They have an even more dramatic effect.

The sheer energy of millions of these burning bullets of glass is also swiftly heating up the planet's atmosphere, so the trees spontaneously combust, as wildfires suddenly ignite and sweep across the landscape. Any dinosaurs who refused to fall in the first firing line of the impact are now roasting, as the temperature becomes hot enough to cook their thick skin and bones.

No more than a quarter of an hour has passed since that ominous and luminous glow loomed in the southeastern sky. And yet most of the dinosaurs are already gone. Ankylosaurs and raptors, duckbills and dome-heads, even rexes and trikes —all dead. The valleys and forests are aflame, and yet some creatures survive. Small lizards and mammals burrow safely beneath the conflagration. Some turtles and crocodiles lurk in the cooling waters, some creatures seek refuge in the freedom of the sky.

Then, the rain stops. During the next hour or so, the air cools and it seems like the worst is over; relative calm has returned to the Cretaceous. But the sky grows blacker from the forest fires. A couple of hours after the first blaze of pure light, the clouds start to roar, and the soot from the fires sweeps up into twisters as an emerging hurricane now blows hard enough across the landscape that the rivers burst their banks.

Sound Garden

With the hurricane comes cacophony. Huge walls of noise louder than any animal has ever heard. No one alive knows that sound travels far more slowly than light. Perhaps it was the faintest notion in the minds of the brightest creatures, who knows. But these sonic booms that now deafen the few surviving dinosaurs are delayed— they left the impact point at the same time as the blazes of pure light. The sound is enough to rupture the eardrums of raptors and send many creatures back into their burrows.

So much for the impact on what is now the western parts of North America. What about the rest of the Cretaceous world? The biblical combination of earthquakes, rock-strewn rain, and hurricane winds are felt in South America and Europe, too, many dinosaurs dying out during those first frenetic and fateful hours.

Elsewhere in the world, the worse is coming to the worst. Huge stretches of the mid-Atlantic coast are being battered and broken apart by tsunamis twice as tall as the Eiffel Tower. The gargantuan wash of the waves is flooding an army of ocean reptiles, including plesiosaurs many miles inland. Over in Asia, in what is now India, volcanoes are vomiting out vast rivers of lava. Meanwhile, "ground zero" of the comet's impact is down in what's now modern-day Mexico. Within a radius of around 600 miles, everything around the Yucatán Peninsula is obliterated, literally vaporized. As the winds die down and the air starts to cool, the ground becomes solid and stable once more. When dusk falls, the glow of the forest fires turns the dark sky to bright amber, as the night draws in on the day when most of the world's dinosaurs died.

Welcome to the Nuclear Winter

Some few dinosaurs survived. Isolated populations will have lingered on, some for weeks and months, some for maybe years. But for those dark days, the Earth was bleak and cold, as the rock and dust of the impact sat in the planet's atmosphere and rubbed out the light of the Sun. The nuclear winter was a climate that only the most robust creatures could endure. The absence of light was a trial for flora, too. As plants depend on sunlight to photosynthesize, their untimely deaths had huge repercussions for fauna "higher up" the food chain. Like a game of Jenga, if the plants are removed, the system collapses and the creatures that depended on the plants die, too. It was the same story in the prehistoric seas. Smaller photosynthesizing plankton perished, as did the bigger plankton and fish that fed upon them, all the way up to the huge reptiles at the peak of the ocean's food pyramid.

In time, the Sun prevailed, eventually shining through and drawing to an end those darkest of days. But there were other problems. The flotsam of the impact hung in the atmosphere and made the rains so acidic that the surface of the planet was scorched. The comet's crashing had also sent trillions of tons of carbon dioxide into the sky. As we all know from today's climate debate, carbon dioxide is one of those greenhouse gases that warms a planet's atmosphere.

And so, quite quickly, in a Cretaceous example of "out of the frying pan and into the fire," a world of nuclear winter evolved into a world of global warming. A starved and ignoble planet emerged, a world where nothing throve. Like the kind of post-apocalyptic world that you see in science fiction films, the once diverse habitat of the Cretaceous had become a strange and desolate landscape. The Age of the Dinosaurs had drawn to a dramatic close.

Explaining the Dinosaur Big Bang

The comet that had crashed into the Earth 66 million years ago had packed some punch. Sure, the comet was only about the size of Sagarmāthā (better known to English readers as Mount Everest), coming in at about six miles across. But when an object that size is also traveling at around 67,000 miles an hour (that's more than 100 times faster than a stealth bomber), you can expect fireworks. The energy of a moving object is known as its "kinetic energy." This is the energy it has by virtue of its sheer mass and movement. Such an energy is calculated as half the product of its mass (the mass of Mount Everest is 357 trillion pounds) and its velocity squared (which is around 4.5 billion). That amount of energy is going to make some bang.

How big a bang did the comet make with the Earth? A bang in the ballpark of a billion nuclear bombs' amount of energy, that's how big. It's hardly surprising that, coming in with that kind of clout, the comet drove 25 miles through the planet's crust and into the mantle below. The crater left by the incoming comet is over 100 miles in diameter and is to be found buried underneath the Yucatán Peninsula in Mexico.

Now that we know the power of the cometary "bomb," we can be a little more forensic about that ill-fated day of extinction. The scene of our particular dinosaur drama was over two thousand miles away from the impact point. Dinosaurs farther south on the continent would have been hit far harder, of course.

Ghost Species

The closer to the comet the creatures were, the greater the catastrophe. For those dinosaurs nearer "ground zero," the blaze of light would have appeared more quickly, the quake of the very Earth would have been more violent, the rain of glassy rock more torrential, and the ensuing heat wave more intense. Any creature within that 600-mile radius of the Yucatán Peninsula would have become instant ghosts, like those poor souls underneath the A-bombs in Hiroshima and Nagasaki who suddenly became shadows on the wall before they could even fall to the ground.

Though too small to earn the title of "planet," comets loom large in literature and folklore. And that's hardly surprising. They seem to simply float high above in the heavens, harmlessly arcing across a fair fraction of the sky. And yet our ancestors also knew the terrible truth: a comet can come crashing into Earth and change the planet irreparably.

How do we explain the terrible events the dinosaurs suffered that day? That first blaze of light was produced when the comet punched into the air around the planet. The comet so savagely compacted the atmosphere beneath it that the air became five times the temperature of the Sun's surface. The second blaze of light occurred when the comet crashed into crust. Because sound travels far more slowly than light, the sonic booms that accompanied the blazes of light moved out from ground zero at much slower speeds and were heard by the dinosaurs after a delay of many hours. With the booms came the hurricanes with winds that blew at speeds of over 600 mph near the source, and even at speeds of several hundred mph as far as two thousand miles away.

Comet Repercussions

The comet catastrophe released a cosmic amount of energy into the planet's rocks. The resulting earthquakes were off the scale, and far more violent than anything we've known in written history. The titanic shock waves in the Earth's crust also set in motion tsunamis that raged across the Atlantic, rooted up building-sized boulders, and lobbed them way inland. In India, the crustal disturbances triggered volcanoes into overdrive, initiating a cycle of eruptions that would last for millennia after the impact, and further complicating the chaos the comet had brought.

All that was solid melted into air. The energy from the impact vaporized not only the incoming comet but also the crust it compacted in the collision. The Herculean amounts of resulting slag were hurtled into the sky. A dark amalgam of vapor, liquid, and rock was jettisoned with such speed that some of the material shot straight out into interplanetary space. Recent research by an international team that have been studying the anatomy of the crater suggests that the comet's trajectory was a "perfect storm." The comet responsible for wiping out three quarters of all species hit the worst possible place on the planet and at the most lethal angle. Not only did the Gulf of Mexico contain huge volumes of sulfur from the mineral gypsum, but at 45 to 60 degrees, the impact was very effective at vaporizing and ejecting debris to high altitude.

Gravity reclaimed most of it. The hot liquid rock chilled down into glassy beads and spears, which relayed huge amounts of heat to the planet's atmosphere, turning the air into a furnace. The incandescent air triggered the forest fires, at least in those continents proximal to Yucatán. There's fossil evidence of the fires in the rocks themselves. The remains of scorched wood and foliage are integral to the strata first laid down after the comet struck. The carbon-heavy soot from the burning forests rose way into the air, along with the rest of the grimy detritus of the conflagration. Too light to be held by gravity, the soot and dust choked up the air currents that flow across the globe. Darkness fell. And the nuclear winter that followed probably finished off most dinosaur life in areas more distal from the comet crater point.

Finally, we see how the volcano scene in *Jurassic World: Fallen Kingdom* is a symbolic rerun of the Cretaceous demise of *all* dinosaurs.

WHAT IS THE "SMOKING GUN" FOR DINOSAUR EXTINCTION?

"To connect the dinosaurs, creatures of interest to everyone but the
veriest dullard, with a spectacular extraterrestrial event like the deluge
of meteors . . . seems a little like one of those plots that a clever publisher
might concoct to guarantee enormous sales. All the [Alvarez] theories
lack is some sex and the involvement of the Royal family and the whole
world would be paying attention to them."

—Ian Warden, *The Canberra Times* (May 20, 1984)

Starring Scientists

The starring character of Michael Crichton's 1990 *Jurassic Park*
novel is Ian Malcolm. A gifted mathematician who specializes
in chaos theory, Malcolm serves as the mouthpiece of Crichton,
who called Malcolm the "ironic commentator" on the dubious
actions of the dinosaur science on Isla Nublar. For example,
witness Malcolm's comment, "Scientists are actually preoccupied
with accomplishment. So, they are focused on whether they *can* do
something. They never stop to ask if they *should* do something."

The starring scientist of dinosaur extinction is Walter Alvarez,
professor in the Earth and Planetary Science department at
the University of California, Berkeley. Alvarez is the "Doctor Alan
Grant" of a real-world dinosaur detective story. It was Alvarez who
first uncovered evidence of the cataclysmic account we detailed

in the previous chapter, but what triggered Alvarez's dinosaur discovery?

It's hard to imagine that one of the biggest impact craters on the planet, over 100 miles wide and 3,000 feet deep, could almost disappear from sight. And yet it did. The live and geologically active planet we know as Earth can cover up its tracks over time. So, when Walter Alvarez first found evidence of the famous "smoking gun" crater, he was almost 6,000 miles away from the Yucatán Peninsula in the Apennine Mountains of Italy, in the Province of Perugia.

The story starts in 1977 in a rocky ravine on the fringes of the medieval village of Gubbio. When looking at the late Cretaceous strata here, Alvarez saw a thin band of clay sitting among the rosy pink dominant limestone of the gorge. Like the crack between piano keys, the clay sat between the Cretaceous rocks below and the Paleocene rocks above. Alvarez didn't know it at the time, but he was looking at a mark of mass extinction, a line that punctuated life and death.

Something else caught the canny eye of Walter Alvarez. He noticed that some of the rock strata he was examining were jam-packed with fossil evidence of many kinds. The remains bore witness to an epoch in deep time that was replete with life. But, above these rocks, and above the clay, Alvarez saw relatively lifeless strata, with few if any of the fossils that sat below. So, Alvarez began to wonder if this line of life and death was indeed a mark of mass extinction—the fingerprint of an event that ruined land and oceans alike, and killed off much flora and fauna, including the dinosaurs.

When in Doubt, Why Not Ask Your Dad?

Could it be, Walter Alvarez thought, that this clay is the key to one of those rare events in deep time when hosts of species suddenly and simultaneously disappeared across the globe? He couldn't be sure. In such circumstances, it is indeed very handy if your father just happens to be a Nobel Prize winner who worked on the historically famous Manhattan Project. The dad in question, Luis Alvarez, was co-discoverer of a number of elementary particles, and was also one of the scientific monitors who flew in an airplane behind Enola Gay to record the effects of fatefully dropping the Little Boy A-Bomb on Hiroshima.

So, there was some pedigree among the Alvarez men. Alvarez Junior conjectured on the chemical composition of the clay. Maybe, he thought, Alvarez Senior had some typically ingenious way of working out how long the clay layer had taken to form. Timing was crucial. If the clay layer formed slowly over millions of years, then the death of the poor dinosaurs, among others, would have been a rather protracted affair. But, if the clay had accumulated quickly, the close of the Cretaceous had been a catastrophic affair.

Now, all planetary scientists know that estimating the amount of time it takes to form a layer of rock is a pretty tricky business, so the Alvarez men came up with a creative solution. They focused on heavy metals, those species in the periodic table of elements that are seldom found in the Earth's crust. One of these heavy elements is iridium, named after the Greek goddess Iris, personification of the rainbow, because of the striking range of colors of its compounds. Iridium is the second-densest metal, part of the platinum group, and the most corrosion-resistant metal.

Up and Atom!

Microscopic amounts of these heavy metals are forever falling to the Earth's surface from extraterrestrial sources. So, the Alvarez men theorized about the heavy elements in that clay layer. If there were only trace amounts of heavy metals, then the layer would have been formed in a relative geological instant. If the concentration of heavy metals was higher, however, then the clay layer would have been laid down over a much more protracted period.

The Alvarez men were amazed with what they found. Not only did they find iridium in the clay layer. They found it in huge abundance. So much iridium, in fact, that it would have taken maybe as much as 100 million years of constant cosmic dusting to deliver that much of the heavy element to the Earth. But that was impossible. After all, the limestone rock layers above and below the clay were reasonably well date-stamped by geologists. The clay could only have been laid down over a few million years at most. The case was becoming more encouraging.

Walter and Luis Alvarez went packing to Denmark. They wanted to be sure that the clay layer in the Gubbio gorge wasn't just some kind of freak exception, and scientific discovery thrives on repeatability, so the Alvarez men went looking for Danish rocks around the same age as those in the Province of Perugia. And, bingo, the rocks at the Cretaceous–Paleogene boundary in Denmark told the same tale. Soon enough, reports came in from other parts of the world; whether they formed on the ocean bed, on land, or in shallower waters, the boundary rocks all divulged the same decisive iridium signature.

Geological Whodunit

The father and son detective team was now faced with a number of geological possibilities. The damage to the dinosaurs could also have been done by slow climate change, flash flooding, or chronic volcanic activity. And yet only one option made scientific sense: as iridium is mega-rare on planet Earth, its origin must surely be extraterrestrial.

A new question was now lit in the nerves of the Alvarez men. Sixty-six million years ago, from the depths of interplanetary space, what was the most likely mode of delivery of an epoch-ending iridium bomb? A supernova perhaps? Unlikely. Besides, no one could doubt the pockmarked surface of the Moon. And for every crater on our less massive lunar neighbor, there are hundreds on the Earth, though our active planet continually covers up its violent past with new crust. An asteroidal or cometary impact was far more likely.

When the father and son team published their dinosaur theory in 1980, there followed a decade of dinosaur-mania. The cometary impact and the dinosaur mass extinction were constantly covered in the media, all the way up to Michael Crichton's 1990 *Jurassic Park* novel and Steven Spielberg's movie adaptation in 1993. The world couldn't get enough of the dinosaurs, and neither could scientists. The Alvarez men's giant impact hypothesis was debated in numerous books and TV programs, along with hundreds of magazine articles and scientific journals. And it was a hypothesis that needed a bevy of different scientific experts: astronomers and geologists, chemists and physicists, paleontologists and ecologists. As a result, the debate began to broaden.

More evidence was uncovered. Not only was the same iridium clay layer found the globe over, but a curious type of quartz was discovered along with it. Quartz is a form of silica, the same as sand. But quartz is a hard, transparent mineral often used in making electronic tech and state-of-the-art clocks and watches. The kind of quartz found alongside the iridium is known as "shocked quartz." Such quartz had only been found in two types of places. One was among the rubble of nuclear bomb tests, where the shock of the nuclear explosion had generated the intense pressures needed to change the quartz lattice. And the other was inside the craters left by meteorites, again formed by the sheer shock waves of the impact event.

Then, there were the tektites and spherules. These are the round and tear-shaped burning bullets of glass cast from the melted detritus of the cometary collision, cooling into glass as gravity pulled them back down to Earth. Evidence of a tsunami was then found in the Gulf of Mexico, deposits that dated way back to the Cretaceous–Paleogene boundary. The case was building fast for a gargantuan event that created titanic earthquakes just at the time when the quartz was being blasted and the spherules were falling.

Jurassic Decade

Finally, as the decade of *Jurassic Park* dawned, the smoking gun crater was finally found. And the reason the crater had almost disappeared from sight? Simply because the geologically active planet Earth had buried the smoking gun under millions of years of Yucatán sediment. Geological work *had* previously been done on the area. But the work was carried out by oil companies, who had kept their findings out of the public domain.

But the hellhole was there. Coming in at over 100 miles wide and 3,000 feet deep, the smoking gun, dubbed the Chicxulub crater, was dated at 66 million years old. In short, it sat at the Cretaceous–Paleogene boundary. The smoking gun is one of this planet's largest crater sites, a measure of just how big that fateful day at the end of the Cretaceous really was. Maybe the biggest comet to crash into the Earth over the last half billion years or so. Like *Jurassic Park*, the dinosaurs were doomed from the start.

WHAT WAS THE GREAT DINOSAUR RUSH?

Ian Malcolm: "God creates dinosaurs. God destroys dinosaurs. God creates man. Man destroys God. Man creates dinosaurs."
Ellie Sattler: "Dinosaurs eat man. Woman inherits the Earth."
—Michael Crichton and David Koepp, screenplay of *Jurassic Park* (1993)

Jurassic Park Dinosaur Rush

In the 2015 movie *Jurassic World*, John Hammond's dream has come true. It is twenty-two years after the events of *Jurassic Park*, and a bankable theme park of cloned dinosaurs has been operating for almost a decade. The Park has indeed captured "the imagination of the entire planet," as Hammond had predicted, and a global rush to see the creatures has meant phenomenal success. And yet the Park's operations manager, Claire Dearing, gives us a hint of what's to come a mere seven minutes into the movie. Park visitors want bigger and better species with more terror and teeth ("Stegosaurus is like an elephant now to kids," she says). This state of affairs explains why indominus rex is transgenically created. Engineered by Park geneticist Dr. Henry Wu, indominus rex is the Park's newest attraction in the hope that the dinosaur rush can continue, and the profits keep rolling in. But there have been dinosaur rushes in real life, too, and the very first rush takes us all the way back to when dinosaurs were first discovered.

In the Shadow of St. Paul's Cathedral

The year 1842 is when dinosaur science really began. That's when British anatomist Richard Owen first coined the word *Dinosauria*, meaning "Terrible Reptile" or "Fearfully Great Reptile." The British Isles is a rather curious place for our story to start. Rarely has its landscape puked up prominent specimens. No trikes or T. rexes. No saurus of either the brachio— or stego—stamp. No velociraptor. And no diplodocus, even though Dippy the diplodocus plaster cast skeleton was proudly presented in the entrance hall of London's Natural History Museum between 1905 and 2017. (Dippy became an iconic representation of the Museum.)

And yet the British Isles holds an important place in the history of dinosaur science. It's the land where the first dinosaur fossil fragments were studied by paleontologists like Owen. It's where the first complete dinosaur skeleton was unearthed. And it's where the name "dinosaur" almost died out a few decades after being christened.

The dinosaur rush began in Victorian London. It may well be, of course, that humans had unknowingly come across the fossilized remains of the ancient beasts long before then. But in the early 1840s, in the center of the city just a few hundred feet north of St. Paul's Cathedral, scientific expert Richard Owen called upon William Devonshire Saull, who was a British businessman, but also a radical activist and keen geologist.

The Discovery of the Dinosaurs

Saull's geological collection boasted the bone of an iguanodon. At least, that's what it had been named a few decades earlier. It was certainly a spine bone from a large ancient creature, but when Richard Owen saw the bone, it made him think. Within weeks, Owen declared that the iguanodon, and a couple of other large ancient creatures that had been uncovered by geologists, was quite unlike anything previously discovered.

Owen dubbed them dinosaurs. Of course, Owen's dramatic announcement in 1842 had a backstory. Much of this dinosaur prehistory also happened in Britain. In 1824, William Buckland, professor of geology at the University of Oxford, had analyzed the fossil of a huge skeleton unearthed in Oxfordshire. He christened it Megalosaurus. And in 1825, a Sussex doctor by the name of Gideon Mantell had found teeth that looked much like gigantic versions of iguana teeth, thus the name iguanodon.

So, Richard Owen's contribution to dinosaur science was not just conjuring up the charismatic name of *Dinosauria*. Owen also realized that creatures such as Megalosaurus, iguanodon, and others shared previously unrealized anatomical features. They weren't just weirdly large reptiles. They belonged to a brand-new species. The dinosaurs had arrived.

But in 1887, a few years before Owen's death, the very idea of dinosaurs was called into question. The British paleontologist Harry Seeley said that dinosaurs fell into two groups. The groups were defined by differences in the pelvis. For example, the "bird-hipped" ornithischians had pelvises like those of modern birds. This kind included beasts such as stegosaurus and iguanodon. Meanwhile, the "lizard-hipped" Saurischians, which embraced giant sauropods such as diplodocus, had pelvises like modern lizards. Seeley believed the two groups were so different that they couldn't have evolved from a single ancestor. "*Dinosauria* has no existence as a natural group of animals," he wrote. "I see no ground for associating these two orders in one group." The idea of the dinosaurs as a single, scientifically valid group was almost dead. It was not until the 1970s that experts were happy once more to entertain the idea that the two groups did, after all, evolve from a single common ancestor. Meanwhile, across the pond, a war had begun.

The Bone Wars

Behind the bones and bling of the Fossil Hall in the Smithsonian National Museum of Natural History is a war that waged over 100 years ago. Benjamin Franklin once said that "All wars are follies, very expensive and very mischievous ones." And this dinosaur war was certainly that. During the paleontology boom of the late 1800s, scientists Edward Drinker Cope and Othniel Charles Marsh went from being friends who named species after one another, to being the frostiest of foes. Indeed, they ultimately ruined one another's careers and even each other's lives. They came for the science and ended up waging war.

The combatants' desperate hunt for fossils led them to the famous Wild West. In the rich bone beds of Nebraska, Colorado, and Wyoming, between 1877 and 1892 the battling paleontologists used their wealth and power to fund their own expeditions and to buy services and dinosaur bones from fossil hunters. By the time the war was over, the men had exhausted their funds in the pursuit of paleontological dominion. The combined expeditions of the pair led to almost 140 new species of dinosaurs being found and cataloged. Indeed, the bone wars resulted in a huge increase in knowledge about prehistoric life, and triggered public interest in dinosaurs, similar to the influence of *Jurassic Park* a century later.

The Foot Soldiers of the New Science

Drinker Cope and Marsh weren't the foot soldiers of the new dinosaur science. That hard work was done by scientific mercenaries such as David Baldwin. Baldwin was sent off to northern New Mexico by Drinker Cope. Get me a fossil, Cope had said. Something gargantuan. Something I can shove in the face of Marsh. At this time, 1881, the bitter feud between Philadelphian Cope and Yale rival Marsh was in its early stages. Yet, neither of the prime combatants had bothered to brave the natural elements of the Wild West.

Native American war parties, such as that under Geronimo, continued in Arizona and New Mexico until 1886. So, instead of fossil hunting themselves, the dinosaur war bosses used a network of hired hands. David Baldwin was one such recruit: an enigmatic maverick who was quite capable of heading out on his mule, deep into the uncharted Badlands for weeks and months on end. Come rain, shine, or the depths of winter, Baldwin eventually emerged with the dinosaur goods.

Baldwin was a turncoat, in the manner of many mercenaries. Having once been in the trusted employ of Marsh, he now hunted dinosaur bones for Drinker Cope. And so, out of the sheer luck of timing, it was Cope who received the fossil prize from Baldwin in the early 1880s. That prize was a set of small and hollow dinosaur bones that Baldwin had eked out of the western deserts. The bones belonged to a creature not known before. A brand-new dinosaur: light in weight, quick as a wink, dog-sized, and sharp-toothed. It was the primitive Triassic dinosaur that Drinker Cope later christened Coelophysis. Pronounced SEE-lo-FY-sis, Coelophysis sits at the base of the family tree of many dinosaurs including T. rex and the ever-popular Raptor family. Coelophysis was also the second dinosaur in space. It traveled there as a specimen skull of a Coelophysis from the Carnegie Museum of Natural History. It was aboard Space Shuttle Endeavor mission STS-89 when it left Earth's atmosphere on January 22, 1998. It was also taken aboard Russian space station Mir before being returned to its home planet.

Wild West Cowboy Bone Hunters

Working men like Baldwin were the storm troopers in the hunt for the biggest dinosaur bones. After all, it was laborers and railwaymen who first started the dinosaur rush. Preferring pale-ontology to hard labor (and who can blame them?), working men reinvented themselves. They tracked the trails of giants, such as the long-necked sauropods like brontosaurus and brachiosaurs, and gargantuan carnivores, such as Allosaurus. Like Baldwin, they became scientific mercenaries, fossil hunters on the payrolls of august institutions such as Yale University.

They were a motley crew. Like Wild West pirates plundering the past in their cowboy hats, Wyatt Earp mustaches, and unruly mops of hair, they hunted fossils for many weeks at a time. They spent whatever free time they had drinking, raiding, and sabotaging each other's sites, fighting, feuding, and shooting at one another. Such was the way in which the bone wars were waged.

The first fossils were surely found by Native American peoples. Having occupied the American continent for well over ten thousand years, they were no doubt the first discoverers of the bones and tracks of dinosaurs on the continent. In some cases, they brought their discoveries to the attention of Europeans. As early as the late 1600s, natives tried to describe the giant monsters that once inhabited the land. For example, Edward Taylor, a minister from Massachusetts, recalled that native people told of such monsters. But he was skeptical of the tales told until 1705, when he saw for himself the huge teeth of a mastodon, which had been found along the banks of the Hudson River in New York.

But the first recorded fossilized bones were discovered during a surveying expedition in 1859, the same year that Charles Darwin published his epoch-defining *On the Origin of Species*. By early 1877, the trickle of dinosaur discoveries had become a flood of fossils. Take the example of railwayman William Reed. Reed was returning from a good hunt. He'd already bagged a pronghorn antelope when he spotted something sticking out of a long ridge known as Como Bluff. The bluff was reasonably close to the railway tracks, in the middle of nowhere in southeast Wyoming. The something that was sticking out were huge, fossilized dinosaur bones. Unbeknownst to Reed, at the same time of his discovery, college student Oramel Lucas was also uncovering huge dinosaur bones a couple of hundred miles to the south in Colorado. And in that same month of March 1877, schoolteacher Arthur Lakes unearthed a cache of fossils down in Denver.

The Dinosaur Rush

And so, the dinosaur rush began. Like the prospecting rushes of people lured by the promise of gold, silver, and other precious metals, the dinosaur rush drew a mob of unruly characters, in this case to the states of Colorado and Wyoming. They came with one mission only: to transform dinosaur fossils into dollars. It didn't take these pirates long to learn which of the head honchos were paying top dollar: Edward Drinker Cope and Othniel Charles Marsh.

Marsh and Drinker Cope treated the Wild West like a theater of war. With each eager telegram from a railwayman or a ranch hand, they saw a chance of finally defeating their famous opponent. So it was that Marsh and Drinker Cope turned that part of the West into a battlefield. Their respective troops behaved like an army of invasion, hoovering up bones and the like as they went, and trying to sabotage the mistrusted enemy. But, as we saw from the case of David Baldwin and many a pirate before him, loyalties in war can be fluid and fickle. Oramel Lucas worked for Drinker Cope. Arthur Lakes marched along under the banner of Marsh. Railwayman William Reed also worked with Marsh, and yet his dinosaur teamsters defected to Drinker Cope. This insanity continued for over a decade. Poaching, pillaging, bribing, and intimidation became the rules of engagement. And yet, when all was done, it was no longer easy to discern losers from winners.

The Spoils of War

The main winner was science itself. By the time a truce was called, or rather by the time both men were ruined, the dinosaur discoveries of the bone wars soon became household names: diplodocus and stegosaurus, apatosaurus and brontosaurus, allosaurus and Ceratosaurus, among others. The downside was that the ethos of the bone wars had resulted in poor scientific practice: fossils hastily ripped from the rock, specimens scantily studied, familiar bones accidentally classed as new species, or fossil bones of the same specimen being incorrectly cataloged as belonging to completely different creatures. Alan Grant would not approve.

As that famous hippie John Lennon once said, "War is over, if you want it." Well, the bone wars *were* over, not necessarily because Marsh and Drinker Cope *wanted* it, but simply because they couldn't afford to continue. The chaos of the dinosaur rush was done. And the world knew a lot more about one of the greatest stories in the history of our planet.

JURASSIC BACKDROP: CONTINENTS AND TECTONICS

"The first concept of continental drift first came to me as far back as 1910, when considering the map of the world, under the direct impression produced by the congruence of the coast lines on either side of the Atlantic. At first, I did not pay attention to the idea because I regarded it as improbable. In the fall of 1911, I came quite accidentally upon a synoptic report in which I learned for the first time of paleontological evidence for a former land bridge between Brazil and Africa. As a result, I undertook a cursory examination of relevant research in the fields of geology and paleontology, and this provided immediately such weighty corroboration that a conviction of the fundamental soundness of the idea took root in my mind."

—Alfred L. Wegener, *The Origins of Continents and Oceans* (1929)

"Plate tectonics: the study of how the surface of the Earth is formed, how the separate pieces of it move, and the effects of this movement."

—*The Cambridge Dictionary*

Isla Nublar

The fictional Central American island of Isla Nublar in the Jurassic Park franchise first appears in Michael Crichton's 1990 novel. It's also the main setting in the movies *Jurassic Park* (1993), *Jurassic World* (2015), and *Jurassic World: Fallen Kingdom* (2018). Both film and novel narratives have the Jurassic theme park situated on Isla Nublar, which is off the west coast of Costa Rica.

And yet not that long ago, geologically speaking, Costa Rica simply didn't exist. Only 50 million years ago, the narrow mountainous strip of land that we now call Costa Rica was part of the ocean floor, along with the rest of the Central American isthmus between Colombia and Guatemala. Now, 50 million years might sound like a long way back in human terms, but in deep time, it's nothing—the dinosaurs had already disappeared 15 million years before.

Supercontinent

The original dinosaurs, those monsters from deep time that were recreated for the Jurassic theme park, evolved on a very different planet indeed. Their world was a connected world. No jigsaw puzzle pieces of ocean and land. Just an immense single supercontinent landmass reaching from pole to pole. Given giant pairs of sneakers and sufficient time, a dinosaur of the day could maybe coastal-walk his way up from the South Pole over the equator to the Arctic Circle. Incidentally, it's a wonderful fact that the Arctic Circle is named after animal life. The word "arctic" comes from *arktikos,* which is Greek for "near the bear," and refers to the constellations Ursa Major (Great Bear) and Ursa Minor (Little Bear), which can be seen in the northern sky.

If our roving, sneakered dinosaur was without a map, he might stray too far inland. Such a mistake might prove rather costly. He may easily find himself perhaps thousands, if not tens of thousands of miles from the nearest beach. Coastal walking is tricky without a coast. If he doesn't stray too far inland, and sticks to the coast, he might consider the ocean. If our creature cared to swim, he could paddle his way around the landmass. But dinosaurs weren't creatures of the sea.

This was the strange world into which the first dinosaurs emerged. Within biological taxonomy, archosaurs are a large group of reptiles, which include all crocodiles, birds, dinosaurs, and flying reptiles known as pterosaurs. Within the archosaurs is a clade, or natural group, known as dinosauromorpha. These so-called dinosauromorphs are reptiles closer to birds than to crocodilians and includes the dinosaurs and some of their close relatives. Birds are the only dinosauromorphs which survive to the present day.

Pangaea

The very first dinosaurs, such as herrerasaurus and eoraptor, evolved from their cat-sized dinosauromorph ancestors between 240 and 230 million years ago. No Pacific Ocean spanned their globe, bounded by the Americas in the east and the continents of Asia and Australia in the west. Their world *had* no North or South America. Nor did it have an Asia, or an Australia. Rather, it had a single supercontinent we call Pangea, surrounded by a global ocean we've named Panthalassa.

Why those names? The name Pangea, or Pangaea, comes from the ancient Greek *pan*, for "entire" or "whole," and "Gaia" for Mother Earth. The idea that the continents once formed a contiguous landmass was first suggested by German geophysicist Alfred Wegener. Wegener was the creator of the theory of continental drift, in which he proposed that, before breaking up and drifting to their current locations, the continents had been a single supercontinent that he termed the "Urkontinent." The name Pangea was used in the 1920 edition of Wegener's *The Origin of Continents and Oceans* (*Die Entstehung der Kontinente und Ozeane*), where he refers to the ancient supercontinent as "the Pangaea of the Carboniferous."

Life on Pangea

As our first dinosaurs were born into this relatively alien world, it's fascinating to imagine how their actual world would contrast with today's "world" of *Jurassic Park*'s Isla Nublar. How did the Triassic Earth compare with our modern world? Well, the Pangea landmass could be thought of as looking like either a huge letter C or, with some artistic imagination, the open jaws of some prehistoric creature. Into those jaws stretched some of the deep waters of the Panthalassa ocean. (The name Panthalassa comes from the ancient Greek *pan*, for "whole," and *thálassa* for "sea."

Pangea was a creation of plate tectonics. According to geological tectonic evidence, the part of Pangea known as Gondwana was assembled by continental collisions in the Late Precambrian, around one billion to 542 million years ago. Gondwana then collided with "North America," "Europe," and "Siberia" to form the supercontinent of Pangea. Huge mountain ranges ran along the landscape, marking the frontiers of collision where smaller plates of crust had crashed and, like puzzle pieces, built the giant continent.

The Climate of Pangea

How did the climate of the Triassic Period compare with that of Isla Nublar? The earliest dinosaurs lived in a world that we could best describe by using the Finnish word "sauna." The atmosphere held more carbon dioxide, which meant more of that famed greenhouse effect—more thermal energy radiated out over Triassic ocean and land. The very geography of Pangea made matters steamier still. For on one side of the world there was land from pole to pole. And in the other hemisphere the open, global ocean. This meant that ocean currents could make their easy way from equator to poles. Water baked at the equator could travel directly to those poles, preventing icecaps from forming.

So, the Triassic world paints a very different picture to *Jurassic Park* and Isla Nublar. Those ocean currents meant that the Antarctic and Arctic were temperate. Summer temperatures at the Triassic poles were like London today (if any Londoners were to complain about this comparison, they need only spend a few moments at today's poles)! During the Triassic, wintry weather was seldom far below freezing. Such were the places where early dinosaurs developed.

Now, you can imagine that if the Triassic poles were balmy, the rest of the globe must have been sweltering. And yet, for reasons of Pangaea geography, Earth did not simply become a desert planet like Arrakis or Tatooine. The Triassic reality was more complicated. As the Pangea landmass was roughly symmetric about the equator, one half of the giant continent was searing in summer as the other half was freezing in winter. The dramatic differential in temperature created fierce currents of air to stream over the equator. And, when the seasons changed, the winds simply switched direction.

Mega-monsoons

Something similar happens in today's world, of course. In Southeast Asia and India, currents of air drive the famous monsoons, the contrast between a dry season and a season of fierce storms and torrential deluge. If you think the June 1993 tropical cyclone that hit Isla Nublar in *Jurassic Park* was bad, think again. Whereas modern monsoons are an Asian phenomenon, Triassic ones were more like mega-monsoons. They were global.

Just as today we see news reports of the raging torrents and sweeping mudslides of modern monsoons, it's easy to imagine Triassic dinosaurs being floored by floodwaters and buried for posterity in an avalanche of mud. And yet, the mega-monsoons had a further, geographical, effect. They effectively separated the supercontinent into micro-climate zones. These zones would have varied temperatures, varying amounts of rainfall, and different severity of mega-monsoonal winds. The equatorial zone was very humid and hot, a veritable tropical hell that makes Isla Nublar seem like Iceland. Next were limited swaths of desert on either side of the equator and stretching out to about 30 degrees of latitude. Desert temperatures would have been over 100°F all year long, nor did the mega-monsoonal rains fall here.

Dinosaur Home in the Mid-lats

The Pangeaic midlatitudes told a different tale. Here, the mega-monsoon was king. Cool, wet, and humid, the midlatitudes were the habitat of herrerasaurus, eoraptor, and the other ischigualasto dinosaurs (dinosaurs named after the ischigualasto formation, a fossiliferous rock formation in northwestern Argentina). The word ischigualasto comes from the defunct Cacán language, spoken by an indigenous group referred to as the Diaguita by the Spanish conquistadors, and means "place where the moon alights."

The early dinosaurs dwelt in the humid belt of southern Pangea. Though the supercontinent was one contiguous landmass, it was a patchwork expanse of hazardous weather and drastic microclimates. Southern Pangea wasn't the most stable of habitats in which to evolve. And yet those early dinosaurs had little choice. Like the rest of the flora and fauna of that time, they were recovering from the huge mass extinction at the end of the Permian Period. The Permian extinction had been the worst extinction event in our planet's history. It's estimated to have wiped out around 95 percent of all marine species, 70 percent of land animals, and was the only known mass extinction of insects. And if that wasn't bad enough, the first dinosaurs now had to deal with the evolutionary battlefield that was Pangea.

Dinosaurs Come Out Fighting

It was by no means a given that the dinosaurs would eventually conquer their habitats, of course. That's not how evolution works. Famous dead ends in evolutionary history include the trilobites, woolly mammoths, and maybe man. Who knows? But back in the days of Pangea, dinosaurs were small and meek, with no guarantee yet that the meek would inherit the Earth.

The dinosaurs were a million miles away from being at the top of their food chain. They shared a strange new world with early mammals, small to medium reptiles, and amphibians. All these creatures sat halfway up their food chain. Peak predators of the chain were the fearsome crocodilian archosaurs. It wasn't going to be easy. At the time of writing, *Jurassic World: Dominion* is around a year and a half away from general release, thanks to the COVID-19 pandemic. It looks as though the new movie's plot has the *Jurassic World* dinosaurs free to roam in our modern world. That's a walk in the park for predators that persevered in Pangea.

Just take one predatory example of what the early dinosaurs had to deal with. The Triassic fossil record shows copious evidence of a beast known to us as metoposaurus. Metoposaurus, meaning "front lizard," skulked around the lakes and rivers of Triassic Pangea, especially the midlatitudes where the dinosaurs roamed, and the subtropical arid zones. Imagine being an eoraptor. You'd have a slender body of around 3 feet in length, and an estimated weight of around 22 pounds. You are lightly built, and quite capable of moving at some speed. All the same, if you're roaming the same shorelines as metoposaurus, you're in enemy territory. Metoposaurus would pounce on anything that strayed too close to the water. It lurked in the shadows, lunging forward with its huge flat head containing hundreds of razor-sharp teeth. Those jaws are hinged like a croc's, so they can snap shut and gulp down unsuspecting dinosaurs like eoraptor.

Dawn of the Modern World

It's often said that the Triassic is the dawn of our modern world. Take the metoposaurus, for example. Like giant salamanders, larger than mere humans, these predators were the evolutionary precursors of toads, frogs, newts, and yes, salamanders. It's a similar story for many of today's identifiable creatures. Their DNA stems back to those early days of Pangea. It's true for turtles, for crocodiles and lizards, and it's also true for us mammals. All these beasts evolved alongside the dinosaurs in the challenging habitats of prehistoric Pangea.

After all, not for nothing was the Permian extinction known as the "Great Dying" or the "Mother of all Extinctions." At first, the Triassic Period was overrun by so-called "disaster species," or "wimpy" species that thrived in disturbed environments. Lystrosaurus was a mammal-like reptile which flourished until the dinosaurs began to have their day. Some scientists estimate that lystrosaurus made up 90 percent of land vertebrates in the early Triassic. This is the only time in Earth history where a single species dominated the land to such an extent. So, the Great Dying was an Armageddon so dramatic that it left an empty evolutionary arena, a theater of life in which all sorts of new beasts could bloom. The first 50 million years or so of the Triassic was like a giant petri dish, a life experiment that altered our world once and for all.

But soon the dinosaurs would have their day. They began to slowly diversify in the southern Pangea in the deep time between 230 and 220 million years ago. Not only were they to be found in the fossiliferous ischigualasto formation, in what's now Argentina, but also in regions of what is now known as India and Brazil, both of which once sat in the temperate humid parts of Pangea. Elsewhere, such as the dry zones close to the equator, dinosaurs were almost completely absent, as far as we can tell. The fossil record shows copious amounts of reptiles and amphibians, but no dinosaurs. Apparently, the first dinosaurs didn't thrive in the searing desert lands.

The Domain of Dinosaurs

Here's the thing you may not expect to read. Dinosaurs did not rule all lands from the outset. They weren't immediately kings of the world. They didn't simply sweep across the supercontinent from the get-go. Rather, the dinosaurs were local. They stayed put, hemmed in by an agreeable climate in their own zones, and a far less likable climate elsewhere.

For many millions of years, the dinosaurs stayed provincial, happy to dwell in those lands in the south of the supercontinent, deterred by the deserts from venturing any farther. In places like the fossiliferous ischigualasto formation, dinosaur species accounted for around 10 to 20 percent of the environment. The rest of the ecosystem was made up of early mammals, other reptile types, and cousins of crocodiles, such as the peak predator saurosuchus.

As time passed, things changed. For one thing, in the humid zone, the dominant herbivores suffered a decline in numbers. It's not known why, but the fall of these plant-eaters meant the fortunes of others improved. New niches in the environment beckoned, so by 225 to 215 million years ago, in the los colorados formation of what is now Argentina, dinosaurs' sauropod ancestors now made up around 30 percent of the ecosystem.

The same was true in the northern humid zones of Pangea, what we now know as Europe. Here, too, long-necked dinosaurs were doing well. A further breakthrough was afoot. About 215 million years ago, in the subtropical deserts of northern Pangea, the first dinosaurs arrived, no longer hemmed in by climate. This ecosystem sat at around 10 degrees north of the equatorial line, in what we would now call the American southwest. Maybe the monsoons changed in nature, making the contrast between the humid and arid zones less marked.

The Supercontinent Splits

The dinosaurs were finally on the move, but so was the world. Around 240 million years ago, the planet's crust began to crack. The cracking actually began before the first true dinosaurs had evolved. Now, there wasn't really much to experience in terms of real drama. This is a slow-burn story, as is most of the history of tectonics. Maybe there were minor earthquakes. Perhaps the odd tremor in that troubled Triassic. But, as the first true dinosaurs such as herrerasaurus and eoraptor emerged, the crustal fracturing persisted, thousands of feet below their "paws."

The supercontinent was splitting, and that eventually brought dramatic change. Tectonic forces are a rich, global amalgam of heat, gravity, and pressure. Given time, these forces are enough to move continents, disappear seas, and raise up mountain ranges. It may help to picture Pangea once more as a giant jigsaw puzzle, prone to being pulled from opposite directions. As the pulling continues, the pieces become thinner, eventually breaking apart. And so it was that over many millions of years, the gradual tug-of-war tore cracks in the supercontinent's surface, as the middle of Pangea's landmass unzipped through its middle.

This tectonic drama just happens to have given us our modern geography. It birthed the Atlantic Ocean. For, as east and west Pangea split, North America said goodbye to western Europe, and South America said adios and ciao to Africa. The Atlantic Ocean was born when seawater flooded in to fill the rift left by the receding land. The Triassic fauna were about to have their lives changed forever, and that included the dinosaurs.

Triassic Park: The Movie

Cue Hollywood blockbuster. That tearing up in the Triassic was no trivial matter. Cue a violent recreation of the continents. Cue an entire landmass lacerating, as lava gushes in glorious technicolor. The planet's outer crust thins and magma from the deep Earth is drawn to the surface through fissures and through volcanoes. When you rend asunder a supercontinent, you can expect Armageddon.

Check out some of the detail of this "Triassic Park" blockbuster. Over half a million years or so (admittedly tricky getting *that* onto two hours of blockbuster film!), plate tectonics orchestrated a quartet of magma tsunamis from the deep Earth. In the days of the age of faith, it was believed that the depths of the Earth housed the locus of the Devil and his legions. In the late Triassic, it must have actually seemed that way! Magma flooded out of the Pangea rift in such biblical quantities that they could have buried the Empire State Building twice over. In total, 3 million square miles of central Pangea were flooded in flowing lava.

If you think the volcano scene in *Jurassic World: Fallen Kingdom* is bad, then imagine our Triassic Park blockbuster. These volcanic explosions are among the largest and most violent planet Earth has ever seen. Big budget stuff. In fact, in the post-COVID economic climate, prohibitively big budget. It was a bad time for mere beasts but, hats off to the dinosaurs, they survived this particular mass extinction. As the lava spewed out onto land, toxic gases bled into the atmosphere and triggered a fast-tracked global warming.

The Triassic extinction, also known as the End-Triassic extinction or the Triassic-Jurassic extinction, was a global extinction event that occurred at the end of the Triassic Period, 252 million to 201 million years ago. It resulted in the demise of some 76 percent of all marine and terrestrial species and about 20 percent of all taxonomic families. And yet this mass extinction also led to the eventual global dominance of the dinosaurs in the Jurassic.

HOW DID DINOSAURS GET SO BIG?

"[My father] started very early to tell me about the world and how interesting it is. When I was a small boy, he used to sit me on his lap and read to me from the [Encyclopedia] Britannica. We would be reading, say, about dinosaurs. It would be talking about the tyrannosaurus rex, and it would say something like, 'This dinosaur is 25 feet high and its head is six feet across.' My father would stop reading and say, 'Now, let's see what that means. That would mean that if he stood in our front yard, he would be tall enough to put his head through our window up here.' (We were on the second floor.) 'But his head would be too wide to fit in the window.' Everything he read to me he would translate as best he could into some reality."

—Richard Feynman, "The Making of a Scientist" (1995)

"You Never Had Control, That's the Illusion"

What is it about the look of the Jurassic Park resort that finally drives home the message of big dinosaur danger to the cinemagoer? What design tricks did the movie production team use to underline the idea of the Park as not only a working and functioning resort, but also hint that things may soon go the way that Ian Malcolm suggests?

The answer is all in the Jurassic Park architecture. Huge amounts of raw concrete are on display. The look of this rough and raw material evokes a primal past. It creates the illusion of an indestructible fortress. Exactly the kind of citadel you'd need if you are mad enough to have an island full of huge and ferocious dinosaurs.

The architecture suggests, what could possibly go wrong? No flimsy tourist resort, this. Just raw resources of height and thickness that seem to boast of their unbreakable nature. And all managed from control rooms, presumably beneath the visitor center, which looks like a nuclear bunker.

The Great Hall of Dinosaurs at Yale's Peabody Museum thankfully looks nothing like a nuclear bunker. It was built in the 1920s to house Yale's enviable dinosaur collection, put together over many years by that motley crew of roughnecks and mercenaries sent west by Ivy League elites. The Great Hall is no Jurassic Park. It has no holograms or flashing digital displays. But it is nonetheless something of a cathedral of science, famous in its own right.

The Age of Reptiles

Running along the entire east wall is a mural that extends over 100 feet long and 16 feet high. Almost five years in creation, the mural was the work of Rudolph Zallinger. Born in Russia, Zallinger was raised in Seattle and became a prominent member of Yale University after painting his murals, gaining him many awards and honors. Little doubt in today's world, Zallinger would work in Jurassic World movies or video games as an artist.

Zallinger so adored dinosaurs that he created his mural homage to the creatures, *The Age of Reptiles*, which then became the significant piece of his work at that time. Some have called *The Age of Reptiles* the "Mona Lisa of Paleontology." And to be sure, they have a point. Zallinger's mural has been a theme for a set of US postage stamps and has been used in all sorts of dinosaur graphics, including features in *Life* magazine. Apart from the Jurassic World franchise, *The Age of Reptiles* is arguably the single most referenced piece of dinosaur artwork ever made.

The mural is a story in time. It's the evolutionary tale of how fish-like creatures emerged from the oceans and onto land. It shows how they colonized their new habitat, thereby branching into reptiles and amphibians. Next, the mural conjures the evolutionary magic of the morphing of the reptiles into both mammal and lizard lines, with the lizards ultimately birthing the dinosaurs.

Toward the mural's finale, a good 60 feet and almost a quarter of a billion years from its beginning, and after a quest through the deep time of primal and alien landscapes, the painting explodes into a cornucopia of dinosaur creatures. All shapes, all sizes, some small and barely seen, but some so gargantuan the brain is seduced into disbelief.

The Biggest of Beasts

The dinosaur section of Zallinger's *Age of Reptiles* shows the majesty of that period in deep time when the dinosaurs had risen to their evolutionary zenith. A colossal brontosaurus sits in a prehistoric swamp, feeding on the ferns and trees that make up the margins of the swamp waters. A brontosaurus would weigh around 30 tons and measure up to 72 feet long. As a useful guide, our largest land animal, the average male African bush elephant, is 10 feet six inches tall at the shoulder and has a body weight of around 5 tons. So, we could say that the brontosaurus was about seven times longer than an African bush elephant and weighed roughly the same as *six* of those elephants! That's some beast.

Stage right is allosaurus, not just tearing into the bloodied flesh of its prey with claws and teeth, but also standing in the very visceral remains of its victim, reminding the viewer of Alfred Lord Tennyson's comment that nature is "red in tooth and claw." Allosaurus is often called the Butcher of the Jurassic. Coming in at 2.5 tons and 30 feet, Allosaurus weaponized its head into a hatchet, with which it hacked its prey to death. This beast could open its jaws incredibly wide, so that when hungry it would attack victims with its mouth open, slashing down at its prey, and piercing skin and muscle with razor-sharp teeth. Many a brontosaurus met its end this way.

Understandably staying well away from Allosaurus, Zallinger's *Age of Reptiles* depicts a quietly ruminating stegosaurus. The Stegosaurus came in at around eight tons and measured 30 feet in length and is pictured fully kitted out in its natural armor of spikes and bony plates, just on the off chance that allosaurus gets too close. Far off in the distance, where the mural's swamp fades into snow-peaked mountains, we spy a second sauropod swooping its long neck down to graze the vegetation on the ground. *The Age of Reptiles* is a magnificent work of art, one which depicts the dinosaurs at the peak of their evolution. And yet, the question remains, how did some of these creatures get so big?

The Sauropods of Skye

Consider Scotland, birthplace of John Hammond in Jurassic Park movie canon. Back in the actual Jurassic, Scotland was a much warmer land. (Scottish people might find this harder to believe than the fact that dinosaurs once roamed there!) A place of swamps and sandy beaches set on an island adrift in the spreading Atlantic Ocean. And, as Pangea continued to break up, Scotland's Isle of Skye was home to not only lizards and crocodiles and beaver-like mammals, but also to generations of sauropods, stomping their way through the Scottish lagoons.

You can easily imagine that many Scottish schoolchildren have often wondered about the mythical Loch Ness Monster. Nessie, as she's known, is of course said in Scottish folklore to inhabit Loch Ness in the Scottish Highlands. Nessie is also said to be large and long necked, sounding curiously like the sauropods that actually *did* inhabit Scotland's past, though admittedly probably not swimming in Loch Ness.

Now imagine that those same Scottish schoolchildren are handed pencil and paper and invited to draw their own mythical beast from deep time. One wonders if the kids would come up with a creation anywhere near what evolution conjured in the sauropods. With very long necks, long tails, relatively small heads, and four thick, pillar-like legs, sauropods are known for their enormous size. The sauropods include the largest animals ever to have lived on land, leaving behind gigantic footprints for future human generations to uncover.

Puzzling Sauropods

In fact, sauropods are so brain-bogglingly big that their discovery was a bit of a puzzle. When the first sauropod bones were unearthed back in the early nineteenth century, it was a time of other fascinating discoveries. Findings such as the carnivore, Megalosaurus, and the herbivore, iguanodon. Compared to these creatures, the Sauropods seemed huge, so scientists mistakenly thought that the sauropod bones belonged to whales. After all, whales are big. Later, of course, not only was the mistake realized, but paleontologists eventually found that sauropods were actually *bigger* than most whales. And this question of size has fascinated scientists ever since.

Just how did sauropods get so huge? The question is one of the most famous puzzles of paleontology. So, let's look at some possible answers. First, we need to estimate how big and heavy we actually mean when we say that sauropods were "huge." And to answer *that* question, paleontologists use dinosaur fossil bones to predict the weight of a dinosaur. Here's what we mean.

Walk the Dinosaur

Huge and heavy animals need stronger legs to carry that weight. Scientists have gathered data on this. They've measured the leg bones of many living animals. No surprise that the thickness of the main leg bone, which supports the animal's weight, is strongly related to the weight of the animal. In short, if paleontologists measure leg-bone thickness, they can then come up with a reasonably accurate body weight for that animal.

Then, there's the digital dynamic. Paleontologists now use 3D digital simulacra of dinosaur skeletons. To these models, the scientists can add skin covering, muscle mass, and even internal organs—all using animation programs. Last, the computer program allows the scientist to calculate the beast's body weight. Photogrammetry is revolutionizing dinosaur science. The associated animation software can be used to make a virtual dinosaur run, jump, and almost dance. (As in the song released in 1987 by the band Was (Not Was), paleontologists literally "Walk the Dinosaur.") Bringing beasts back to life in this way, such techniques can also be used for dinosaur animation in television or film.

Sauropods Were Huge

Paleontologists have used experimental techniques to show that sauropods were indeed huge. Take Plateosaurus, for example. This primitive proto-Sauropod was knocking about in the Triassic. And, even back then, Plateosaurus was able to weigh up to 3 tons. That's about half the weight of an average African elephant. By the time Pangea began to fragment into the puzzle pieces of the continents, sauropods proper got even bigger.

The Sauropods of the Isle of Skye in Scotland that we mentioned earlier amassed to around 10 or 20 tons. By the time of the Jurassic, citing monsters such as brachiosaurus and brontosaurus, beasts would weigh more than 30 tons. But to truly tip the scale to gargantuan, we have to look to the Cretaceous. Titanic species such as Argentinosaurus and dreadnoughtus weighed more than 50 tons. That's about seven times as heavy as a T. rex, about twelve times as heavy as a Hippopotamus, and roughly the same weight as a Boeing 737. So, how were these beasts able to amass to such a size, especially when compared to the rest of the fauna produced by evolution?

Supersize Me

Well, for one thing, sauropods would need to eat a lot of food! When paleontologists think about sauropod size and the nutrition available from Jurassic foods, they reckon that a brontosaurus, for example, would need to chomp down on about 100 pounds of flora every day. Twigs, stems, leaves, the lot. And that means they'd need to collect and digest all this food. Then, there's growth. If you're going to get huge, you'd better start early. After all, if you take your growth rate *too* slowly, it might take you 100 years to get gargantuan! And by *that* time, you might have been totaled by a falling tree, done in by a deadly disease, or butchered by some badass predator.

Then, there's breathe and build. To maximize sauropod size, a creature would have to breathe with great efficacy, in order for them to derive sufficient oxygen for the powerful metabolisms of their huge bodies. Build-wise, the beasts would need to be built with a strong and sturdy skeleton, and not too bulky, as anything excessive would hamper locomotion. And last, they'd need to get rid of any surplus body heat. It's all too easy in hot weather for a big beast to simply overheat, keel over, and collapse, often fatally.

So, how did Sauropods manage all of this? Perhaps the physical conditions of deep time, Triassic, Jurassic, and Cretaceous were slightly or maybe markedly different. Take gravity, for example. Maybe it was weaker. Maybe the big beasts were able to grow and move more easily in those long-gone days. Or perhaps there was a greater percentage of oxygen in the Earth's atmosphere. More oxygen would enable the titanic sauropods to prosper more easily. And yet, there's little evidence that gravity was very different in the days of the dinosaurs, and oxygen levels may actually have been slightly lower.

Sauropods- "R"-Us

That leaves us with the conclusion that just maybe it was down to the sauropods themselves. Something about the Sauropods broke the evolutionary mold. The size constraints that held back reptiles and mammals, amphibians, and other dinosaur species didn't seem to apply to the sauropod's body plan. Nature's genius, through the Triassic and early Jurassic, appears to have evolved in the sauropod body plan a creature well adapted to prosper at titanic sizes.

Let's begin with the elongated neck. That trademark neck—long, rangy, and majestic—began its development back in the Triassic with the proto-sauropods and just got longer over time, as species after species evolved additional vertebrae. The elongated neck enabled the sauropods to reach those parts of the flora that other herbivores just couldn't reach. Exclusive access to a new food source.

The neck had other advantages. A sauropod could plonk itself in one place and from that anchored spot simply range its neck where it chose, cherry picking the local produce at will. Minimum energy used for maximum grub up. And conserving energy put them one step ahead of their competitors, as well as their necks, allowing them to eat the diet needed to accumulate mass.

Growth Spurts

Sauropod growth must have been something to behold. Their first advantage was in their genes. They'd gotten a head start from their ancestors back in the early Triassic. Back then, the so-called dinosauromorphs had faster metabolisms, quicker growth spurts, and busier lifestyles than many of the reptiles and amphibians of the day. They certainly weren't slothful. Neither did it take them an age to develop into adults, like the crocodile or the iguana. The same became true for all the dinosauromorphs' dinosaur descendants.

The second advantage was that Sauropods found their fast-track. Scientific research shows that most sauropods grew from the size of a guinea pig to the dimensions of a dragon in around 35 years. That's hardly any time at all for such an amazing transformation.

Sauropods inherited another trait from their Triassic forebears: powerful lungs of high efficiency. Sauropod lungs are like those of birds, and very different indeed from the lungs of you and me. We mammals have very basic lungs that simply take in oxygen and give out carbon dioxide. Birds have so-called unidirectional lungs. Here, air flow is in one direction only. And oxygen is drawn out during inhalation and exhalation. This type of lung is efficient as it extracts oxygen at every breath in, and every breath out. That's evolution for you. The bird-type lung is enabled by a set of air sacs linked to the lung. These then store some of the oxygen-rich air drawn in during inhalation, which is then passed from the lung to the air sacs when exhaling.

How can scientists tell that Sauropods possessed such a bird-type lung? Because the uncovered bones of their chest chamber have large cavities, known as pneumatic fenestrae, indicating that the air sacs ran deep into the body. Modern species of bird have the same structures. And such structures are only made by air sacs. And so, to sum up the third Sauropod advantage, they had extra-efficient lungs that enabled sufficient oxygen uptake to power their metabolisms at titanic size. Incidentally, theropod dinosaurs also had bird-type lungs. This may have been a factor in enabling tyrannosaurs to get as big as they did, and why stegosaurs, horned dinosaurs, and duck-billed dinosaurs were not able to grow as gargantuan as the sauropods.

Air-Sac Secret

Yet, air sacs have a further, unintended trait. Apart from air storage per se, air sacs invade bone structure, and in so doing they lighten the load of the skeleton. Air sacs hollow out parts of the skeletal structure, making a still strong but less heavy bone load, in a similar way that an air-filled soccer ball is lighter than a boulder of the same size.

So, what about those long and rangy sauropod necks? How come the creatures can hold out their necks without falling over? The answer also lies in the air sacs. Sauropod backbones are so bejeweled by air sacs that their vertebrae are like the honeycombs of the honeybee—lightweight but resilient. In short, air sacs gift our sauropods evolutionary advantage number four: they enable the creatures to possess a skeleton that's strong, light, and mobile.

The fifth and final factor of the sauropods' evolutionary advantage is their ability to lose their surplus body heat. Yet again, the air sacs helped. Along with the lung, the air sacs provide a large surface area for dissipating excess body heat. Every blistering breath of the beast would be cooled down by the creature's biological air-con.

Recipe for a Supergiant

That's how the supergiant dinosaurs were made. Should the evolving sauropods have lacked any of the evolutionary advantages—rangy neck, speedy growth rate, proficient lung, lightweight skeleton, or super-cooling air sacs—they probably wouldn't have turned into the titans they became. That's evolution for you. Take every step at a time, build the beast in the correct order, and when the creature is finally created in those long-gone days of the Jurassic, a new breed is born. A breed so huge, they earn the right to become known as titans, sweeping the globe and remaining dominant and majestic for another hundred million years.

WHAT PREDATORS POSE THE BIGGEST THREAT IN THE JURASSIC SERIES?

Your tour car pulls up next to a tall, electrified fence. You have arrived at the Tyrannosaur Paddock. The sign reads DANGER 10,000 VOLTS. You're filling with anticipation. On the other side of the fence, you can see lush green bushes and palm trees backing up into a tropical rain forest. Farther back, the vegetation-covered mountains loom with their summits partially obscured by low-lying clouds. Somewhere in this scene lurks a king, and you want to see it.

To tempt the tyrant lizard king, a live goat is raised into position as bait. Surely this isn't legal, but you're frozen, eyes locked on the goat. Nothing is happening, though. Then, the dinosaur expert sitting in the car next to you remarks, "T. rex doesn't wanna be fed, he wants to hunt, you can't just suppress 65 million years of gut instinct."

Although you're aware that the T. rex is actually a female, the statement is compelling. Surely the T. rex wouldn't turn its nose up at a free meal? You start thinking about it while you wait. Would a T. rex hunt or scavenge? What would typically be on their menu in the wild? What do carnivorous dinosaurs munch on? And most importantly, which dinosaurs should you be afraid of?

Mesozoic Mealtimes

Remember that scene in Jurassic park when Alan Grant, Lex, and Tim are up a tree and get approached by a brachiosaur? Tim points out that this particular dinosaur is a vegesaurus, much to Lex's relief. She later applies this rationale to inquire upon the feeding preference of the flocking gallimimuses, asking, "Are those meat eating, er . . . meatasauruses?"

Reflecting back to science lessons at school, you probably already know that the terms normally used for these descriptions are herbivores and carnivores, and for creatures that eat a bit of anything, the term omnivore is used. However, when it comes to being stuck in the wild with these dinosaurs, the main thing you'd probably want to know is whether you're on its menu or not. This is really a matter of understanding which part of the food chain you're in, in comparison to that particular animal.

A food chain represents how energy and nutrients are transmitted from one life form to the next and is part of a wider set of interactions, known as a food web. Each stage on a food chain or web represents a different trophic level. Trophic basically means "nutrition related," and the trophic levels range from one to five. Each successive level relies on the level below for its nutritional needs.

At the base (level 1) are plants, known as producers, because they produce their own food (i.e., autotrophic) through photosynthesis. At the higher trophic levels are all the heterotrophic creatures, meaning that they survive by consuming other life forms and as such are known as consumers. All animals are consumers.

The lowest group of consumers are the plant-eating herbivores (level 2), also known as primary consumers. Then, there's the meat-eating secondary consumers (level 3) made up of creatures that feed on herbivores. Then, tertiary consumers (level 4) are carnivores that feed on other carnivores. Any carnivores that actively hunt other animals for food are known as predators and if they themselves have no natural predators they're regarded as an apex predator (level 4 or 5). Lions and sharks are apex predators, as were tyrannosaurus rex and Spinosaurus.

Some have suggested that T. rex may have been a scavenger as opposed to a hunter, an idea promoted by *Jurassic Park* scientific consultant Jack Horner. Based on the knowledge of T. rex at the time, he pointed out that the large size, small arms, strong sense of smell, and inability to run fast were possible indicators of T. rex being an obligate scavenger (i.e., it only scavenges). Although he also later said that he held this stance to be contrary and to get people arguing. He actually believed that T. rex was a facultative scavenger (i.e., it scavenged on occasion), but would otherwise rely on predation. This is observed in the vast majority of living vertebrates that are known to scavenge.

In the movies, T. rex is portrayed as a predator, a view generally accepted by paleontologists and supported by evidence. For example, it's been shown that they could actually run fast enough to catch certain prey, while their relatively short arms wouldn't have prevented their powerful jaws and strong ziphodont teeth (blade-shaped and serrated along the edges) from grabbing onto prey. However, the most compelling evidence comes from fossilized remains that show signs of attack from a dinosaur matching T. rex's profile. For example, visible bite marks have been found in the brow horn and neck frill of various triceratopses, while in 2013 a T. rex tooth was found embedded in the tail of an edmontosaurus. In the latter case, the herbivore lived on to heal its wounds after the ambush, supporting the idea that T. rex had attacked it in life, rather than just biting into its lifeless carcass.

Now, if the young Lex or Tim were to come across any of these great beasts, how could they tell a meatasaurus from a vegesaurus?

Recognizing a Carnivore

As a consequence of natural selection, creatures evolve to occupy certain roles, or niches, within their environment. From generation to generation this process slowly chisels them into forms with particular features that aid survival in their habitat. As such, by assessing a creature's physical form, it's possible to infer how its body may have functioned within its environment. For example, by comparing fossilized teeth and bones with modern animals it's possible to see trends related to how they may have acquired and consumed food. This, of course, isn't the only available evidence. Paleontologists also make use of computer models, fossilized gut contents, and even fossilized poop to get insight into a dinosaur's feeding habits.

Carnivores have certain traits that make them stick out, depending on their particular habitat and feeding preference. Carnivorous dinosaurs were generally bipedal theropods, with active physiques that helped them to hunt prey. They also featured pointy teeth, known as ziphodont teeth. Typically, their claws were also sharp and recurved (i.e., curved backward), which helped in catching and holding on to their prey.

A well-known *Jurassic Park* carnivore featuring these prominent claws is the velociraptor. In real life, they were only turkey sized, but the velociraptors appearing in the movies were actually modeled on their larger relative deinonychus, which lived earlier in the Cretaceous. Also, while Velociraptor fossils have only been found in Mongolia, deinonychus fossils were actually found in Montana, where Alan Grant is based in the movies.

It was studies on deinonychus by real-life paleontologist John Ostrom that helped to push the notion that dinosaurs were active, agile creatures and potentially "warm-blooded." Ostrom gave deinonychus its name, meaning "terrible claw," and by identifying similarities between the feathered, bird-like archaeopteryx and the nifty deinonychus, he was able to push the notion that birds likely evolved from the bipedal and (mostly) meat-eating theropod dinosaurs.

The fact that deinonychus appeared agile implied it had a high metabolism to suit, which suggests that theropods, at least, may have been warm-blooded. A warm-blooded or more correctly endothermic creature would require more food to sustain its basal (base level) metabolism. As a carnivore, the act of hunting and catching prey would also burn substantial energy, requiring more food again. In general, endothermic dinosaurs needed more food than if they were ectothermic (so-called "cold-blooded").

The possibility of endothermic dinosaurs was something popularized by Ostrom's student, Robert Bakker, who is mentioned by Tim Murphy in *Jurassic Park* and also parodied by the character Robert Burke in *The Lost World*. You might remember Robert Burke as the guy who gets snatched by the T. rex after a snake crawls onto him and he freaks out. Next to the T. rex, the unfortunate human was just another creature occupying a lower trophic level. But what creatures would these and similar prehistoric predators typically have tucked into?

Up the Food Chain

There were various prey available to carnivores, from fish and insects to mammals and reptiles. As is observed in modern animals, dinosaurs had particular features that increased their ability to obtain their main types of food.

Occupying the lowest carnivore trophic level were the herbivore hunters, considered as secondary consumers. A few are believed to have been insectivores and would have typically been small enough for an insect diet to sustain them. An example is albertonykus borealis, an alvarezsaurid dinosaur with a long snout, small teeth, short and powerful limbs, and claws suited to digging. As its forelimbs were too short to burrow, it's believed it used them for searching out termites in rotting wood.

Another secondary consumer and probably the most memorable small creature in *Jurassic Park* is Compsognathus or "Compies" as Eric Kirby called them in *Jurassic Park III*. You might also remember them from the beginning of *The Lost World* when a bunch of them attack a young girl and then later maul the injured and quite annoying hunter Dieter Stark. There's currently no evidence that they actually hunted in packs or attacked animals as big as humans, but fossilized Compies were found in Europe with the remains of small lizards in their stomachs, which they appear to have swallowed whole. Their ability to catch these nimble lizards indicates that the Compies were quite fast and agile.

Real-life Compies of the Late Jurassic lived alongside the smaller archaeopteryx. Archaeopteryx is a member of the bird-like paraves, a group containing some of the smallest non-avian dinosaurs. The paraves also contain the ancestors of the aves (modern birds) as well as the group that the Cretaceous velociraptors are a member of, called deinonychosauria (terrible clawed lizards). Deinonychosaurians can be recognized by the distinctive raised sickled claw on their second toe, believed to have been used to pin down and hold prey while they ate them alive. However, as opposed to the movie description of them slashing their victims, it's now thought that it was likely their jaws rather than their claws that acted as the main tool for the kill.

In 2018, scientists carried out microwear studies and finite element analyses of tyrannosaur and deinonychus family teeth, revealing that many of them used a "puncture-and-pull" feeding strategy. Furthermore, dinosaurs like deinonychus were well suited to handling struggling prey, as their teeth appeared to be good at handling stresses applied from multiple biting angles. This means that larger animals than them (that presented a bigger struggle) were fair game and that their powerful bites would not have been out of sync with the capabilities of their teeth.

One such beast suggested as prey for deinonychus is a herbivore called tenontosaurus, many of which have been found along with the remains of deinonychus. A fully grown tenontosaurus weighed about a ton. Due to their size, it's been proposed that the tenontosaurus were either preyed upon as juveniles, or else deinonychus attacked in groups in order to bring one down.

Predatory dinosaurs weren't limited to land-based meals either. Some appear to have been fond of fish. These fish eaters (or piscivores) are recognized by their particularly slender jaws and straight, conical teeth used for spearing and holding on to fish. Crocodilians share these characteristics, as do the dinosaurs known as spinosaurids, leading to the view that they likely lived a piscivorous lifestyle. More evidence of this lifestyle came from a spinosaurid called baryonyx, which had fish scales found in its stomach. Although, it also contained the bones of a young herbivore called iguanodon. So, although spinosaurids were well adapted for dining on fish, it appears that they may have also eaten any other animals they came across, including humans fleeing from a crashed plane.

One familiar spinosaurid, called Spinosaurus, is considered one of the largest carnivorous dinosaurs and likely didn't have many predators itself. As such, it was an apex predator, occupying the highest trophic level in its food web, one of many spread across the continents of the Mesozoic.

Apex Predators

In *The Lost World*, a Spinosaurus is the main villain and can be seen rampaging across Isla Sornar. When it gets confronted by a T. rex, it promptly dispatches of the tyrannosaur using its huge claws, while the T. rex is clamped between its jaws. These two dinosaurs did not exist at the same time so would never have come into contact, and it should also be noted that T. rex only roamed in the ancient western United States, while Spinosaurus was native to Northern Africa. Nonetheless, both were considered the top predators of their time, filling the spot at the top of the food web in their respective environments.

Another top predator we see in the series is the Late Jurassic allosaurus, featured in *Fallen Kingdom* and *Battle of Big Rock*. This 30-foot-long, 3,300 pound predator was known to have feasted on stegosaurs and some sauropods, and would have been one of the top predators in its habitat, which it shared with Ceratosaurus. Ceratosaurus was the red-faced, horned dinosaur that appears briefly in *Jurassic Park III*. It approaches and then retreats from the characters who are next to a river, digging through piles of Spinosaurus dung to find a buried satellite phone. The presence of the Ceratosaurus in this habitat is consistent with theories that suggest it may have favored smaller prey and semiaquatic food and locations. Although, this four-fingered beast lived about 50 million years before Spinosaurus, so they would have never actually shared their habitat.

Another *Fallen Kingdom* prominent predator is the bull-horned carnotaurus, which also features in the animated spinoff series *Camp Cretaceous*. It dominated Late Cretaceous South America around the same time that tyrannosaurs roamed the north. Carnotaurus is thought to have attacked large creatures like sauropods, but we instead meet it threatening our protagonists just before being taken out by the much larger tyrannosaurus rex.

T. rex was a ferocious carnivore as revealed by studies of its teeth, jaws, droppings, and body structure. It used its teeth to puncture and grab flesh, then rip it away, sometimes with a vigorous shake of its head. (Imagine a dog with a toy.) Often, it would lose a tooth or two while feeding, evidence of which has been found strewn across the environment near or among the fossilized remains of its prey. Its banana-shaped teeth weren't uniform in length and varied slightly throughout its jaw, analogous to how we have incisors, canine, and molar teeth.

T. rex's teeth took up to two years to grow and were constantly being replaced, as is the case for all toothed dinosaurs, although at different rates. For instance, in 2019 it was estimated that the teeth of the Madagascan predator majungasaurus were replaced every couple of months, while for Allosaurus and Ceratosaurus it took just over three months. Unlike any other known dinosaur, though, T. rex's teeth were strong enough to crunch through bones, a fact backed up by computer and biomechanical studies of its skull strength and biting force. Evidence of this durophagy (ability to eat hard things like shell and bone) is also present in analyses of its fossilized feces, where bones of its prey have been found.

The Biggest Threat

Those were some of the biggest threats in the Jurassic series, but if we ever found ourselves adrift in a Jurassic world, the threats wouldn't just be from the apex predators. We'd be food for most of the predators around our size and bigger. Although, within our particular trophic level, there'd be many other creatures that we could inevitably out-compete due to our intelligence. But where would we best be located on a Jurassic series food web?

Well, we're omnivorous predators that can also be preyed upon by other animals, so we're definitely not at the top of the trophic tower. A study in 2013 set out to find our position in the global food web by calculating the human trophic level (HTL) based on global diets (which is actually about 80 percent plant produce and 20 percent meats). The study found the global HTL to be 2.21, that is, between level 2 (primary consumers) and level 3 (secondary consumers), adding that "humans are similar to anchovy or pigs and cannot be considered apex predators." So, in the global food web, our position is pretty much equivalent to pizza topping: mere fast food for higher level predators, which in the context of the movies is kind of how many humans end up.

In reality, the situation is a bit more complicated because, given the right tools, humans can also prey upon any of the other so-called apex predators, elevating us above them. According to behavioral ecologist Liana Zanette from Canada's Western University, humans should be regarded as a "super predator" because we kill large carnivores and "middle-of-the-food-chain" animals at a much larger rate than those animals would usually be killed. So, as the only super predator in existence, we would actually occupy a lone spot at the very top of the food chain.

Then again, that's all academic. Imagine standing between two apex predators like the lion facing off against the T. rex at the end of *Fallen Kingdom*. A "wise human" positioned between a "king of beasts" and a "tyrant lizard king." In such a scenario, being a super predator would probably be the last thing going through your mind . . . well, maybe that and the T. rex's teeth!

HOW DO YOU FEED A HERD OF HUNGRY HERBIVORES?

Lex: "Shh. Shh. Don't let the monsters come over here."

Dr. Alan Grant: "They're not monsters, Lex. They're just animals. And these are herbivores."

Tim: "That means they only eat vegetables, but for you I think they'd make an exception."

—*Jurassic Park* (1993)

We all gotta eat and the consumers occupying the lowest regions of the food web are the herbivores. In terms of biomass (i.e., the mass of life forms in a given area), the numbers increase as you move down the web toward lower trophic levels. Carnivores are dependent on a large population of herbivores and in turn herbivores are supported by an even larger amount of plant material, so the more food you have at the bottom of the web, the more animals you can support at the top.

At the start of *Jurassic World*, Gray Mitchell says there are fourteen herbivore and six carnivore species, apparently feeding on 50 tons of food a week. Most of this would have been devoured by the herbivores, many of which are known to have congregated in herds. Feeding these herbivores isn't just a matter of quantity, though. The type and quality of the vegetation has a major impact on what can feast on it and also affects how much vegetation will be needed on the island.

So, how do you cater for a herd of hungry herbivores?

Herbivory

Herbivores are primary consumers with representatives from all of the major dinosaur groups. The long-necked sauropods (like diplodocus and brachiosaurus) are herbivores, as are nearly all ornithischians. The ornithischians include the thyreophorans (armored dinosaurs like stegosaurus and ankylosaurus), ornithopods (such as the hadrosaur parasaurolophus), and marginocephalians (like triceratops or the dome-headed pachycephalosaurs). Even the third major dinosaur group, the typically meat-eating theropods, had herbivores among their ranks, such as the long-clawed therizinosaurs.

Don't worry if those names were a bit of a mouthful, the main point is that most dinosaur groups were actually herbivores. For a place like Jurassic Park, the food for these herbivores would ideally come from vegetation growing on the island, as is the case with many zoos, but some parts of their diet are often topped up from other sources to ensure that they get adequate nutrition. Isla Nublar actually had a herbivore feeding compound dedicated to its plant-eating population.

Plants provide herbivores with all of their nutrients and energy. Energy content is measured in calories, with one gram of fat providing nine calories of energy, while protein and carbohydrates each provide four calories of energy per gram. Meats are generally high in protein and fat, whereas plants are high in carbohydrates (so one kilogram of meat can provide more energy than one kilogram of vegetation). This means that herbivores need to eat more food than carnivores to get the same number of calories.

The number of calories an animal needs is largely related to its metabolism, with endothermic animals tending to have higher metabolisms than ectotherms because they need extra calories to maintain their body temperature. The jury's still out on the precise setup for dinosaurs but it very likely varied between groups as well as different ages and sizes. The largest dinosaurs were all herbivores, and their size would have made them less susceptible to external changes to their core temperature, a phenomenon called gigantothermy. This means that the larger groups could have been as active as endotherms while benefiting from the smaller food intake of ectotherms.

Another thing to consider is that herbivores in captivity tend to eat less than their wild or free counterparts, as they don't have to travel huge distances to forage for food, which is kept topped up by the Jurassic Park staff. To maintain adequate diets the herbivore handlers would have had to observe each new species to see how much it needed to eat and then supplement their diets as necessary. Different foods have different nutritional levels, though, and in extant herbivores, there are different digestive systems to extract adequate nutrition from the plants.

Gut Issues

Plants, like all life forms, are made of cells that are bounded by a flexible membrane. However, plant cells also have an additional hard outer layer called the cell wall. The cell wall contains a substance called cellulose, which provides strength to the plant but also makes it tougher to eat and digest. In fact, vertebrates can't directly digest cellulose at all, which is why the mostly cellulose husks of corn kernels can pass straight through our system, emerging undigested in our poop. Different plant parts contain different amounts of cellulose, though.

Animals use special enzymes to break down proteins, fats, and carbohydrates, such as sugar, starch, and cellulose. If an animal lacks the necessary enzymes, its body can't break those substances down. This applies to certain toxins, too. The enzyme that breaks down cellulose is called cellulase, and although it's not made by most animals, it is made by particular microbes, which herbivorous mammals like cows and horses have inside their digestive systems.

This means that herbivores can extract more calories from plants than non-herbivores.

Herbivores have different ways of breaking down food, though. For example, ruminants, such as cows, pass their food through multiple stomach chambers that contain food-fermenting bacteria, while also repeatedly regurgitating their food for extra chewing (i.e., ruminating). Chewing helps to break up the plant cells and coats the food with saliva, which begins to break down starch and fat while making the food more slippery to swallow. After passing through multiple stomach chambers, the food passes to another chamber, called the cecum, where it encounters more cellulose-busting bacteria. Because their digestive system extracts more nutrients from their food, ruminants can eat lower quality vegetation than other herbivores.

Horses do things a bit differently. They are monogastric, meaning that they only have a single stomach. However, they also have a larger cecum and colon, both of which contain cellulase-producing microbes to break down cellulose. In the horse, the microbial fermentation occurs after the main stomach section, so it's called a hindgut fermenter. It's thought that most herbivorous dinosaurs were likely hindgut fermenters.

It's also likely that some dinosaurs even swallowed stones to help them break down food, similar to birds today. Birds lack teeth to chew their food, so instead they swallow stones (known as gastroliths) into their stomachs to help pulverize the food internally. Gastroliths have been associated with some sauropods, a ceratopsian called psittacosaurus, and some toothless theropods such as gallimimus (which may have actually been an omnivore rather than a strict herbivore).

It's not enough to just provide a general set of plants for the herbivores; some require other things, like gastroliths, to support their eating. The vegetation also needs to be something that their guts can actually obtain sufficient nutrients from, while not being toxic to them. Diets also change with age and size, so any feeding strategies must be able to cater to these changes throughout the life of an animal.

Size Matters

There were many small herbivorous dinosaurs, but a good deal were megaherbivores weighing over 2,200 pounds, with some larger sauropods coming in at around 80 tons. According to an observation known as the Jarman-Bell Principle, the bigger an herbivore's body mass, the less selective it is of food and the poorer quality its diet is. Bigger animals tend to be less picky eaters but can pack in more bulk to gain sufficient nutrition. Generally, animals that aren't so picky about their food and just whack the whole thing in are known as bulk feeders.

On the other hand, some animals are selective feeders who tend to go for choice parts of a plant or even stick to particular species. Regarding this feeding strategy in a megaherbivore, it was noted by South African researchers in 2012 that, "Despite their narrow selection for plant species, large size coupled with hindgut digestion enables elephants to exploit a wide range of plant parts, including fibrous stems, bark, and roots." So, their large size and guts allowed them to still indulge in a wide range of vegetation.

Mesozoic megaherbivores were avid eaters and relied on bulky plants that could regenerate quickly, such as conifers and ginkgoes. Although, all plants weren't created equal, and experiments on horsetails and conifers such as Araucaria (e.g., monkey puzzle trees) have shown that when fermented, most of them release more energy than many ferns, cycads, and podocarp conifers. Additionally, in 2019, a team lead by Carole Gee from the University of Bonn reported that horsetail was likely an important "superfood" for young sauropods, helping with their fast growth in early stages of life.

Generally, as young dinosaurs grew, they occupied different niches depending on their size and specific features. For example, a new hatchling might compete for resources with small mammals and reptiles but as a juvenile may occupy a niche alongside mid-sized herbivores like pachycephalosaurs. The young may not just be smaller copies of the adults, either. Sometimes different parts of their anatomy grow at different rates, meaning that their growth is allometric. This is seen in sauropods, and the related growth changes can provide indications about their diets at different stages of life.

For instance, recent research into a small diplodocid skull collected in Montana in 2010 revealed that the youngster (nicknamed Andrew) had peg teeth at the front and spatula-like teeth at the rear of the jaws. This is different from the adults that typically have one type of teeth and implies that their diet may have been more varied at that stage of life compared to when they get older. (This would be useful knowledge to have when feeding the baby dinosaurs of Jurassic World that are seen running around a fenced-off pen for the amusement of visiting kids.)

A Space of Their Own

The variously sized herbivores had to share their environment, and on a small tropical island, this would put an obvious strain on the amount of vegetation available. This is problematic, as pointed out by plant scientist Susannah Lydon:

I also wonder a lot about the sheer biomass of plant material required to feed a herd of Mesozoic megaherbivores. Even if we assume that their metabolism is not endothermic, you're still talking vast amounts of food. Your island would have to be big to allow for regrowth after deforestation when the herd came through.

These concerns aren't limited to the fictional Jurassic Park islands, either. The Dinosaur Park Formation in Alberta, Canada, contains one of the most diverse collections of fossilized herbivores, who all shared the same habitat within a time span of roughly one million years. Assemblages like this puzzled paleontologists as to how the herbivores all managed to survive on the available vegetation within their environment.

The answer came down to the fact that the animals all had their own specific morphologies and adaptations, linked to their food supplies and environment. Basically, they occupied the same vicinity, but they didn't all eat the same types of vegetation, so they weren't competing for the same resources. This species-related division of resources is known as niche partitioning and it means that many different creatures can share the same habitat without exhausting their main food supply. This can be seen in large extant herbivore populations, such as on the African savanna where evidence has shown that in addition to having different food preferences, the animals also feed at different times of the day or in different parts of their habitat.

On a tropical island such as Isla Sorna, these habitats might include grass savannas, swamps, volcanic mountains, tropical and temperate rain forests, and littoral (near to the shore) forests. The forests also have associated layers such as the canopy, understory, shrub layer, herb layer, and forest floor, each of which can support particular groups of herbivores who feed at different heights, known as feeding height stratification.

Feeding Heights

In the Dinosaur Park Formation mentioned earlier, most of the discovered herbivores could only feed at less than three feet due to their physical limitations. This includes ceratopsids and ankylosaurs but generally applies to all thyreophorans and marginocephalians. So, if these animals were put into the same area, they would have all been accessing foods such as ferns, shrubs, and horsetails within the low-lying herb layer.

Some dinosaurs may have also started incorporating fruit into their diet in the Late Cretaceous. For example, a small Australian ankylosaur was found with "chopped plant fragments, seeds, and fruit-like objects" in its stomach. Typical of many herbivores, ankylosaurs had leaf-shaped teeth, suited more to chopping, so they didn't do much chewing. Instead, they developed a sizable gut that enabled longer digestion of the devoured plant matter. These guys likely grazed the lower-level plants and based on their jaws, it's thought that they could have handled quite tough plants, too.

Various herbivores weren't limited to just one feeding height, either. For example, hadrosaurs could feed on vegetation up to almost 7 feet in height while in a quadrupedal stance but also had the option to adopt a bipedal stance to reach vegetation 16 feet up. This meant they could also feed among the mid layers and devour shrubs, bark, and low-growing trees, including cycads, gingkoes, and smaller conifers. Evidence has also shown that some hadrosaurs dined on rotting wood, in which the tough plant cell walls had broken down into a form that could be more easily digested.

Reaching up a bit higher, long-necked sauropods like brachiosaurs and titanosaurs were tall enough to browse the treetops of 100-foot-tall conifers, gingkoes, and monkey puzzle trees (Araucaria Araucana). Brachiosaurs had longer forelimbs than hind limbs and a more vertically held neck, which are both traits that assist in high browsing, although there's still some uncertainty about whether some sauropods actually held their heads more horizontally. However, it does appear that some sauropods were well adapted to rearing up onto their hind legs to increase reach. Although, due to the brachiosaurus's center of gravity being more forward than other sauropods, it's believed that this probably didn't apply to them in the way it was portrayed in *Jurassic Park*.

Feeding the Hungry Herd

There's lots to consider when supplying a suitable diet to the vast range of herbivore diets in the Jurassic Parks. Any food wouldn't do. The dinosaurs would need plants that they could digest and that depends on how their guts work. Some could process lower quality food but eat more of it, while others could dine on high-quality vegetation, relatively packed with nutrients.

You'd also need a variety of plant species of different sizes to cater to the different niches occupied by the dinosaurs. This would mean covering the range from shrubs up to tall trees, including leafy and bushy plants as well as some that bore fruits. There would have to be enough of it to cater to the different numbers of herbivores in each feeding area, and as it got used up, there would need to be fast-growing plants that could quickly replace any lost vegetation.

The amount of food needed would actually depend on the dinosaurs' metabolism and how active they were, but you'd have to be prepared to provide supplemental food to boost their intake. This might involve shipping in specially grown produce or growing it on the island and processing it to feed some of them more directly. This could be especially useful for very young dinosaurs.

It would also make sense to only host animals that naturally settled into different niches. This would allow them to all occupy the same habitat without competing, maximizing the use of space on the island and increasing the chance of supporting a varied population of herbivores and megaherbivores. Having said that, the Jurassic Park designers may have been better off just choosing smaller dinosaurs to inhabit the island, or maybe just genetically engineering them like that. The island could support more animals that way. Paleontologist Stephen Brusatte agrees that this could be a fantastic idea, adding that, "Yeah, cloning dinosaur species that were actually evolved to fit into a smaller, more resource-limited habitat like an island would probably be heartier in a Jurassic Park scenario."

This isn't just a thought-up scenario, either. Nature has already set a precedent, as discovered by Hungarian paleontologist Franz Nopcsa, who in the early 1900s found evidence of dwarf dinosaurs that inhabited an ancient Romanian island. The animals evolved this way due to the limited resources in their environment, which would also be the case on the Jurassic Park Islands. It's called insular dwarfism. So, ecosystems involving smaller dinosaurs are totally possible and, although maybe not as breathtaking as the bigger dinosaurs, would be a much more sustainable way of populating a small island with dinosaurs, while ensuring you could always feed the hungry herds.

WHAT'S WITH THE GIANT DINOSAUR POOP?

A sick triceratops lies on the ground, barely moving and surrounded by inquisitive humans. One of them lies on its belly as it takes huge breaths that move the human up and down. Not that it's got any energy to care much about that. Something's up with it and its designated keeper is baffled, but ancient plant specialist Ellie Sattler offers some insight into its condition.

Judging by its dilated pupils she suspects the herbivore may have eaten something disagreeable from the local plant life and, to be sure, she wants to see the dinosaur's droppings.

Cue Ian Malcom.

"That is one big pile of shit!"

It may not have been immediately apparent, but there's a whole lot of analysis that can be applied to the not-so-humble dinosaur poop portrayed in this scene. So, if you were ever mystified by the mammoth mound of dung in Jurassic Park, hold your nose, because we're about to go deep.

Fecal Facts

The amount an animal poops depends on its daily activity, metabolism, and the type of food it eats. When animals eat food, they digest many of the proteins, fats, and carbohydrates, but dietary fiber (which is undigestible plant-based carbohydrates) just passes straight through. Triceratops was an herbivore, so its diet was high in carbohydrates and fiber, meaning that a bigger proportion of plant-based meals passed through its system and came out undigested in its poop.

Carnivore poop is quite different in that the animals they eat contain lots of fat and proteins that are more fully digestible (although bones, feathers, and fur can still pass straight through the digestive system). Also, since you never asked, herbivore droppings are generally less smelly than a carnivore's. So, Ellie Sattler piling through triceratops poop hits a stink factor of about three out of ten, compared to Alan Grant piling through Spinosaurus poop in *Jurassic Park III*, which I'd give a nine out of ten.

No one really knows exactly how much a 30-thousand-pound triceratops may have pooped, but (at the very least) it wouldn't excrete more than it originally ingested. As a very rough comparison, male African elephants are the largest land-based herbivores, weighing as much as 14 thousand pounds. In the wild, they spend about 80 percent of their day devouring more than 330 pounds of vegetation. Some have even been quoted as having almost twice that intake.

As a general rule, compared to smaller animals, larger species consume a smaller proportion of their body mass in food. But even though elephants only digest about 45 percent of the vegetation they ingest, they still manage to defecate a mass of matter equivalent to what they eat. This is because fresh dung can contain up to 90 percent moisture, and elephants drink more than 150 liters per day. So, let's assume that the triceratops eats and poops like a large elephant, eating just over 2 percent of its body mass in food. In this extremely simplified case, the largest triceratops would eat more than 600 pounds per day, leaving almost as much in droppings. Whether they'd drop this load all in one go, though, is pretty unlikely.

Dropping a Load

African elephants poop up to ten times a day, averaging about 33 pounds per load. However, the biggest animal that ever existed, the blue whale, can release two hundred liters (about 440 pounds) in one go, though the circumstances and consistency are totally different. In 2018, a team at Georgia Tech studied the poop practices of different-sized mammals from mice to elephants. They found that regardless of the size of the animal and volume of poop, the evacuation time averaged twelve seconds (give or take seven seconds), eased along by the presence of a slippery mucus lining that coats the exiting excrement. Whether this holds for reptiles or birds is unknown, but if dinosaurs had a similar relationship, it means that whatever they were dropping was laid within twenty seconds.

While curling one out, many animals don't squat like humans, dogs, and cats do, as that increases the risk of predators catching them with their pants down. The scent of their poop can also make them a target to predators and more dominant members of the species, which is why fewer dominant cats tend to bury theirs. On the whole, though, some animals defecate indiscriminately wherever and whenever the urge takes them, while others are a bit more targeted with their turds.

Pooping is more than just emptying their bowels, though. Consider the dung-dropping behavior of rhinos. They're one of many mammals that defecate communally, in dung piles called middens. According to a 2017 study from South Africa, these middens may act as a way for rhinos to communicate territory or reproductive state via dung odor. If a similar thing were done by triceratops, the pile could have been the result of many different poopetrators visiting the same drop zone repetitively. Although, to be fair, the dung in these middens is generally spread out rather than piled in a mountainous heap.

Speaking of which, most megaherbivores do their doos while standing rather than squatting, and the maximum height of an animal's droppings is generally limited by the height of its rear orifice, which for dinosaurs is the cloacal opening. The cloaca sits at the base of their tail where it meets the hip, so it's located at less than half the height of triceratops. Triceratops were almost 10 feet high, so their departure lounge would sit at about 5 feet high. It's hard to imagine any feasible situation where a creature could excrete something higher than its own rear orifice without flying, projectile pooping, or indulging in some kind of bizarre Jurassic Park Jenga. Since a triceratops couldn't produce a pile that large or high, especially in one sitting, what's with "the big pile of shit," as Malcolm puts it?

The Droppings Dilemma

In the movie, we get the initial impression that the giant dung heap belonged to the sick triceratops. (Although, to be fair, it's actually the smaller pile that Dr. Sattler gets elbow deep into.) There's also another pile just in shot but it's not certain whether this selection of stools is meant to depict droppings from a single animal deposited in multiple sittings, or else the droppings of multiple animals. Let's consider some possibilities.

The large pile is almost as tall as Ian Malcolm, played by the 6' 4" tall actor Jeff Goldblum. Using him as a reference, the large pile can be estimated at just under 6 feet high, with a base about 13 feet across. Assuming the pile is roughly cone shaped, sloping at 49 degrees off vertical at the apex, its volume is about 261 cubic feet, or equivalent to a cube whose sides are almost 6.4 feet in length. If we now multiply that volume by the average density of the dung, we have a very rough estimate of its mass.

The density of dung varies, depending on its total solid content, but in the 2018 Georgia Tech study mentioned above, the researchers also noted that carnivore poops are "sinkers" while herbivore poops tend to be "floaters." As things have to be less dense than water to float in it, herbivore poop must be less dense than water. So, by using the density of water, we can work out the weight that the pile won't exceed, which is a total of 16,314 pounds. This is just over half the mass of a large adult triceratops.

Now, the large pile was definitely not from one animal unless the trike had been perching its poo on the same pile for almost four weeks straight. So, maybe the most rational solution is that the wardens gathered up all that poop into one big pile to clear it up, likely for use as fertilizer. As such, the smaller two-foot-high pile is probably a better representation of a lone triceratops's droppings.

We'll treat the pile very broadly as a cone, but with a steeper slope that's 35 degrees off vertical. The actual pile is flattened off at the top, but if it went to a point, the apex would be a little over 2.5 feet high, in which case the mass of the mess would be no more than 660 pounds. This isn't too far off the amount we calculated for the daily load of a large triceratops, so if for some reason the trike let out its daily deposit in one load, this small pile seems like a fair representation of what we'd see. Now, glancing back over to the mammoth mound again, there's another option we could consider.

Miscellaneous Mound Makers

At a long shot, maybe the larger dung pile wasn't from a triceratops at all and belonged to another dinosaur with a bigger appetite and higher exit hole. Available candidates on the island include the marginally higher Parasaurolophus or the sufficiently tall T. rex or brachiosaurs. The problem with these is that the only T. rex on the island was still confined to her enclosure at that point, while Parasaurolophus and brachiosaur shared a separate paddock from the Trikes. Although, considering the sheer amount of waste, a sauropod like brachiosaur may have been a candidate if it had somehow wandered into the triceratops paddock.

This is admittedly pushing the limits of likelihood, but it does at least provide an opportunity to mention the following wonderful demonstration. In the 2018 BBC series *Deadly Dinosaurs with Steve Backshall*, the British naturalist remarks that one of the biggest sauropods, Argentinosaurus, may have evacuated up to 3,300 pounds per day and subsequently sets out to visualize the impact of dropping such a ludicrous load. To do this, they brought in 3,300 pounds of horse manure, which was incidentally a pile only a couple of feet high and spread over an area maybe 4 x 7 feet. After shoveling it onto a front loader truck, they lifted it to the height of an Argentinosaurus's "poop shoot," and dumped the whole load onto a caravan, whose roof completely collapsed under the weight.

They remark that it's unknown whether the poop would have been deposited in one lump or as smaller separate droppings, but they opted for the former, which was much more dramatic. Even more so if you imagine a smaller dinosaur passing by the business end at the time. In any case, the amount they used was more than five times greater than the triceratops's daily load calculated above, but still five times less than the potential maximum mass of the more humungous dung pile.

Now, if you're feeling pooped out after all those overly detailed exercises in speculative scatology, spare a thought for those paleontologists who have dedicated their days to the study of actual dinosaur poop.

Coprology: The Scientific Study of Poop

Scientists can gain a lot of information about animals by investigating their dung, and if that dung has long since fossilized, it's referred to as a coprolite. It's remarkable that even after millions of years a dinosaur's poop can help fill in some gaps about a dinosaur's lifestyle and environment.

We've already noted some of the differences between herbivore and carnivore poop, but some differences are also visible in the coprolites. Herbivore coprolites are actually pretty rare, with most coprolites coming from carnivores. A carnivore's diet contains more bones and protein, which, after going through the animal's digestive system, produces excretions that are more likely to become fossilized. In particular, if a subsequent sample is found containing high levels of phosphorous, it's a sign that the culprit had a protein rich diet, meaning a carnivore probably laid the log.

While herbivores typically spend more time digesting food, carnivores have higher metabolisms with food passing through their digestive systems faster, coming out partially digested or even undigested. Undigested remains can indicate what the creature had dined on and sometimes how it devoured its meal, such as whether it swallowed it whole or chewed it to bits. So, in *Jurassic Park III*, when the satellite phone is found intact after passing through the Spinosaurus's digestive system, it would suggest that the creature didn't chew its food much before swallowing it.

In the movie, the presence of the phone was a good indication that the dung was deposited by the Spinosaurus, but scientists examining the excrement could have found various other poop inclusions to identify the owner. For example, if they found plant fibers and bits of broken twigs, it suggests it was an herbivore or omnivore's dung, while teeth, fur, or feathers indicate a carnivore laid the poop. Then, broken bones are a sign of a bone-munching dinosaur, while fish bones, scales, or shellfish reveal that the pooper may have dined in or near a water source. All of these clues can and have been found by paleontologists investigating coprolites. Some even reveal details of the animal's digestion.

Traces of preserved muscle were discovered in one coprolite studied by world-renowned fossilized-poop specialist Karen Chin from the University of Colorado. The presence of the intact muscle fragment meant that it didn't have time to be digested by the carnivore that ate it, pointing toward the devourer having a high metabolism. Karen Chin was also responsible for studying the biggest-ever dinosaur dung found yet, which was over 17 inches long and thought to belong to a tyrannosaur. They know this because the size and shape of a coprolite can indicate how big the animal was that produced it, and the only large carnivores (in that area) capable of producing the types of remains found in that coprolite was a tyrannosaur.

Dr. Chin has also found coprolite traces of snails and burrows caused by dung beetles. These creatures may have had a role in the handling of dinosaur droppings, like their current role. People often overlook the organisms (such as insects, fungi, and bacteria) that move and convert the waste at the bottom end of ecosystems, but they are a key part of modern ecosystems, and it would have most certainly been the same when dinosaurs lived.

Dung and Dusted

We've looked at a big pile of turd and rolled up our sleeves to get a better understanding of what that giant poop pile was all about. We've seen how investigating dung is a serious matter, practiced by scientists who literally know their shit. Hopefully, we've come out the other end a little more informed about a matter that would usually make us turn up our noses.

We can't underestimate the value of poop, whether it's a tiny coprolite or a mammoth mountain of mess. Ellie Sattler knew that the poop potentially had a lot to reveal, and the fact that she went straight to the reasonably sized pile shows that she was under no illusion about who could have laid what and when.

So, what's with the giant dinosaur poop? Well, it's just the agglomerated remains of an herbivore's diet, rich in fiber and low in odor, and maybe even riddled with dung beetles. Considering the mammoth effort put into transforming the island, it stands to reason that this man-made heap was destined for collection by waste management, to be redistributed for the growth and maintenance of the island's vast plant population.

MEET THE JURASSIC FAMILY: AT HOME WITH THE TYRANNOSAURS

"Tyrannosaurs weren't always tyrants. For millions of years, the ancestors of the regal T. rex were relegated to second-class predator status while a different dinosaur dynasty ruled over what is now North America: towering allosaurs. But the allosaurs went extinct during the late Cretaceous, allowing tyrannosaurs to seize the throne and then evolve into large killing machines like T. rex."

—Nicholas St. Fleur, "Tiny Tyrannosaur Hints at How T. Rex Became King," *The New York Times* (2020)

"While the most famous of the species is the mighty T. rex, tyrannosaurs came in all shapes and sizes, and their history extends over 100 million years. Despite their final demise during one of Earth's biggest mass extinction events, tyrannosaurs live on both in popular imagination and even through to their present-day bird cousins. Tyrannosaur research is one of the hottest areas in paleontology—several species have been described in just the past decade—and exciting new discoveries are regularly re-drawing the family tree."

—National Museum of Scotland Tyrannosaurs Exhibition (May 2020)

Apex Predators

Tyrannosaurs make plenty of appearances in the Jurassic World franchise. As an apex predator that can reach over 40 feet in length, tyrannosaurus was the largest species of the tyrannosaurids.

Tyrannosaurus, of course, was meant to be the star attraction at John Hammond's original *Jurassic Park*. During the Isla Nublar Incident in 1993, a female T. rex escaped from her compound and went feral on the island. There were eight tyrannosaurs on Isla Sorna, the island initially used by InGen as a cloning station for their dinosaurs. A family unit of an adult male, an adult female, and an infant were discovered during the Isla Sorna Incident in 1997. The male was ensnared and transferred to San Diego, where it was to be used in the failed theme park, which Hammond had intended before procuring Isla Nublar. But the adult male swiftly escaped, wreaking havoc on America's finest city, before being returned to Site B. Three years later, another male Tyrannosaur was encountered on Isla Sorna, where it was finally seen off by a Spinosaurus.

The Jurassic World franchise is very keen to portray a pack of tyrannosaurs at the peak of the food chain. But where did they come from? What's the backstory on the tyrannosaurs, these so-called "tyrant dinosaurs"? Well, over the last couple of decades, paleontologists have unearthed around 20 new tyrannosaur species. They've been uncovered the globe over, from the south of China and the snowy tundra of the Arctic to the arid landscape of the Gobi Desert and the seaward cliffs of south England. These fossil finds have allowed researchers to paint a better picture of this Jurassic family. But what was life like at home with the Tyrannosaurs?

The Long Road to Dominance: Kileskus

Tyrannosaurs were primeval predators whose lineage traces back more than 100 million years before T. rex. The first tyrannosaurs were relatively unimpressive when compared to T. rex in those prosperous days of the Middle Jurassic. Early tyrannosaurs were second-class, human-sized carnivores who stayed that way for 80 million years. They lived in the marginal shadows of more potent predators such as allosaurus, then the carcharodontosaurs, but eventually the Tyrannosaurs rose to the peak of the food chain, ruling the Jurassic world in the last 20 million years of the dinosaurs' dominion.

To get a better grasp of tyrannosaur evolution, consider kileskus. Kileskus is known from partial remains found in 2010 in a Middle Jurassic Formation of Sharypovsky District in Siberia. Given its dissimilarity to Isla Nublar and the like, Siberia is admittedly not the first place you imagine when conjuring up dinosaur scenarios, and yet new fossils are being found the world over, including places like the remoter parts of Russia.

Currently considered the oldest tyrannosaur, kileskus is evidence that tyrannosaurs got an early evolutionary leg up the ladder of life. Kileskus was unearthed in rocks from the middle part of the Jurassic, around 170 million years ago. With nowhere near the dramatic impact of a T. rex, kileskus weighed little more than 100 pounds, and was roughly seven or eight feet long, with most of that length belonging to its skinny tail. Kileskus would have come up to your waist, and its predatory position in the environment would have been something like a wolf or a jackal. A relatively insubstantial hunter, which used its speed to stalk smaller prey. How do we know? Well, for one thing, in the Sharypovsky District in Siberia where kileskus was found, the quarry is also busting out with fossils of mammals, small lizards, turtles, and salamanders. In short, kileskus-kill. Nonetheless, as kileskus isn't exactly chomping on rangy-necked sauropods and bus-sized stegosaurs, it's pretty hard to imagine how this meek little creature could have given way to a monster like T. rex.

The Crown Dragon

How do we know that one begot the other? How do we know kileskus is an ancestor of T. rex and, given its hunting habits and its size, how can we even be sure that kileskus is a tyrannosaur? Well, it's all down to other fossil finds. The partial remains of kileskus are very similar in nature to another small carnivore from the Middle

Jurassic. And complete skeletons of this other meat eater, one adult and one teen, were carefully uncovered in western China.

The Chinese team of scientists named their discovery guanlong, meaning "crown dragon," after the fact that the bone crest sitting along the top of the critter's skull is punctured by a series of holes. Guanlong's crest-bone is somewhat like the ostentatious tail of the male peacock. Or maybe like the outrageous dorsal fin on stethacanthus, the extinct genus that resembled modern sharks up to a point, save for its fin—the shape of an ironing board—that appears to have been part of a courtship display, as it is found in the males only.

The logic of paleontologists goes like this. Guanlong has many traits that it shares with only T. rex and other large tyrannosaurs. For instance, it has fused nasal bones at the top of its snout, along with many other anatomical features probably too tedious to list here! Yet, these evolutionary tell-tale signs are enough to conclude that guanlong is a tyrannosaur. And, as kileskus shares plenty of traits with guanlong, kileskus is also likely to be an early tyrannosaur.

Meet the Tyrannosaurs

How did those initial tyrannosaurs behave? What did they look like and how did they fit their prehistoric niche? Using the anatomical data found by those paleontologists in China, we can tell that guanlong was a loose-limbed little critter, weighing in at about 150 pounds, with a long tail that balanced its lean body. Guanlong was swift in the hunt. It boasted a gob-full of sharp predatory teeth, with long, strong arms and deadly claws—a world away from T. rex's infamous wimpy limbs.

Yet, guanlong wasn't king of the hill. Consider its contemporary carnivore contenders for the crown of peak predator: monolophosaurus, coming in at over 15 feet in length, and the wonderfully named sinraptor, twice as long as monolophosaurus, weighing a solid ton, and a close cousin of the fearsome allosaurus. So, guanlong had second-class predator status. An anonymous node in a food network domineered by fitter dinosaurs. And the same would have been true for kileskus.

So, meet the tyrannosaurs. They're a family of theropod dinosaurs spread all over the world during the roughly 50 million years from the Middle Jurassic and way into the Cretaceous Period, from around 170 to 120 million years ago, and whose fossils have been found in North America, Asia, Britain, and likely Australia.

The tyrannosaurs were survivors. They abided through the changes in climate and habitats that did in dinosaurs such as the mighty allosaurus, stegosaurs, and sauropods at the Jurassic Cretaceous threshold. The tyrannosaurs roamed freely as they lived at a time when the landmass of Pangea was still tearing up into puzzle pieces. Lithe and swift, those tyrannosaurs crossed the land bridges into the new continents with consummate ease. They found their niche, a place in the scheme of things, even though second-class predators, where they could watch and wait.

Tyrannosaurs Get Feathers

Tyrannosaurs gradually got bigger. In the early part of the Cretaceous, around 125 million years ago, some may have even reached sizes of 10 to 12 feet and weighed 1,000 pounds or so. If you were mad enough to attempt it, you could even have tried saddling them up and seeing how long you stayed on the back of such beasts. (Except, of course, there were no humans around to actually *try* such a task, but you get my drift!)

A little over a decade ago, a team of Chinese paleontologists found fossils in the northeast of their country. The fossils belonged to what's since become known as sinotyrannus, meaning "Chinese tyrant." Sinotyrannus was a big tyrannosaur. Best estimates by scientists put this beast at about 30 feet in length and around a ton in weight. If the estimates are accurate, this tyrannosaur would have been ten times the length of Guanlong. With fossils around 125 million years old, sinotyrannus became the oldest big-bodied tyrannosaur found to date, but the evidence that sinotyrannus left behind was sketchy.

Luckily, though, Chinese scientists found yutyrannus. Yutyrannus, or "feathered tyrant," was a tyrannosaur that lived in the early Cretaceous in what is now called northeastern China. Yutyrannus was like sinotyrannus. It too was a tyrannosaur. It had the showy bone crest. And it was big. Yutyrannus was also about 30 feet long. But this was no longer an estimate. It was hard evidence based on the three complete yutyrannus skeletons that had been unearthed.

So, the speculation turned out to be true. There were indeed bigger tyrannosaurs in the early Cretaceous. Incidentally, the other remarkable feature of yutyrannus was the immaculate preservation of the soft tissue. In the vast majority of cases, of course, there are little fossilized remains in terms of skin, muscle, and internal organs of a creature. They decay, or are eaten away, long before the creature becomes fossilized in the first place. But, luckily, the three complete yutyrannus skeletons were buried very soon after a volcanic explosion. Their soft tissue was preserved.

Yutyrannus was feathered. Not the kind of feathers you find on modern birds, the quill-pen type they use in the Harry Potter stories. The feathers on yutyrannus were more prosaic. They presented as strands of slim filaments, about six inches in length, and looking more like hair. This evidence of yutyrannus helped paleontologists confirm that tyrannosaurs were among the feathered dinosaurs. Feathers used for display, rather than for flight.

Tyrannosaurs Get Big

Now, feathers are fine, of course. But, for now, we need to know how the Tyrannosaurs grew into the terrific film-star size of T. rex and Co. When did they start to become the monsters we know from the movies, with huge skulls, almost 40 feet in length, and with two tons of mass bounding around on thunder-thighs of sheer muscle?

The truth is that this kind of movie-star monster appears to have surfaced around 82 million years ago. They were found in Asia. They were found in North America. And soon they were found all over the globe. Fossil evidence seems to suggest that the turning point came in the mid-Cretaceous. Sometime between 110 and 82 million years ago, the world went from the global presence of medium-sized tyrannosaurs, with a few larger beasts such as yutyrannus, to the monster tyrannosaurs, which reigned only in Asia and North America. No smaller tyrannosaurs remained.

This is one of the most startling revolutions in the whole history of dinosaur science. And, sadly, there's not a lot of fossil evidence to help us with the mystery. As (bad) luck would have it, the mid-Cretaceous is not a good time for paleontologists. Nor was it a good time to be a dinosaur. For a long period of around 25 million years, not much fossil evidence was laid down. What we *do* find from that period are hikes in temperature, deep seas bereft of oxygen, and ocean levels turbulently fluctuating. It could be that spasms of volcanism spewed huge amounts of carbon dioxide into the air that would have caused a rapid greenhouse effect, poisoning the planet.

Tyrannosaurs Survive and Thrive

Yet, the tyrannosaurs survived. Around 82 million years ago, when the fossil evidence begins to pick up, the huge Tyrannosaurs had made their way to the status of peak predator in Asia and North America. Now was their heyday. For the last 20 million years of the Cretaceous Period, the tyrannosaurs dominated in the lands where they roamed. On floodplain and in forest, on lakeshore and in deep desert, and in river valleys narrow and wide, all across Asia and North America the Tyrannosaurs ruled.

Now, we begin to see the tyrannosaurs that we know and love. The gargantuan head. The ripped frame. The muscular legs. The long powerful tail. The bite so fierce that it pierced and merely munched through the pathetic bones of their prey. The growth rate so accelerated that they daily gained 5 pounds in weight as teens. Their turbocharged lives meant that nary a single specimen is found older than 30 at time of death.

The tyrannosaurs showed notable diversity. Paleontologists have unearthed around twenty different late-Cretaceoun Tyrannosaurs. The world in which they ruled was very different than the one from which their ancestors evolved. When kileskus, guanlong, and yutyrannus chased down their victims, the landmass of Pangea had only recently begun to break up into its continental puzzle pieces, so migration was a pretty easy affair for the early tyrannosaurs.

By the time of the monster tyrannosaurs, continental drift had configured the globe in much the same way as it looks today, but there were some dramatic deviations. For example, sea levels meant that North America was divided in two. A huge ocean-way ran down from the Arctic all the way to the Gulf of Mexico. Europe, too, was limited compared to today. The continent was merely a sporadic set of small islands.

The tyrannosaurs inherited a patchwork planet. Groups of different dinosaurs prospered in different parts, and should a dinosaur reign as king in one region, there's a good reason he may not reign in another: without a boat there was no way of reaching that distant shore. (And dinosaurs weren't known for their sailing.) Monster tyrannosaurs never reigned in Europe, but in Asia and North America, T. rex became king of all.

WAS T. REX KING OF THE JURASSIC WORLD?

"Of all the organisms scientists have found in the fossil record, tyrannosaurus rex is the most prominent ambassador for paleontology. No dinosaur hall is complete without at least some fragment of the tyrant dinosaur, and almost anything about the dinosaur is sure to get press coverage. We simply can't get enough of old T. rex."
—Brian Switek, "Tyrannosaurus: Hyena of the Cretaceous" (2011)

"Note: In the rare situation a mega-tsunami washes a T. rex into your path, you won't be carrying a weapon large enough to hurt it. If it's intent on eating you, it will eat you. However, you will be killed by the coolest dinosaur ever. Most people go their whole lives without ever seeing a T. rex in person. Do you know how lucky you are?"
—Andrew Shaffer, How to Survive a Sharknado (2014)

Resurrected Infamy

Conjure a picture of T. rex before a bunch of young kids and they'll know right away what beast it is. The characteristic body plan. The enormous signature head and jaws. The ripped frame, and the pathetic forelimbs and stubby fingers. Made infamous by the movies, T. rex was a celebrity dinosaur decades before *Jurassic Park*. American paleontologist Robert Bakker said that T. rex was the "most popular dinosaur among people of all ages, all cultures, and all nationalities." And so it seems to have been. In 1905, Henry Fairfield Osborn, then President of the American Museum of

Natural History, declared the tyrannosaurus the greatest hunter ever to have strutted the Earth.

T. rex enjoyed glowing reviews from the start. On December 30, 1905, the *New York Times* described the dinosaur variably as "the absolute warlord of the earth," a "royal man-eater of the jungle," the "king of all kings in the domain of animal life," and "the most formidable fighting animal of which there is any record whatever!" Little wonder that early moviemakers wanted to get this monster "prize fighter of antiquity" and "Last of the Great Reptiles" into the new cinemas.

T. Rex at the Pictures

T. rex soon had a starring role in many major movies. The 1918 film *The Ghost of Slumber Mountain* was not just among the first of such films, it was also most likely the first movie to show T. rex battling it out against triceratops. Another movie milestone was the cinematic adaptation of Arthur Conan Doyle's *The Lost World*, where stop-motion technology was put to dramatic and spectacular effect. For the times, anyway.

Then came *King Kong*. The celebrated stop-motion animator Willis O'Brien, who worked on *Slumber Mountain* and *Lost World*, created the incredible special effects for this hugely famous 1933 monster movie. *King Kong* featured a climactic fight between the giant ape and a T. rex. The Japanese movie monster Godzilla, whose franchise spawned thirty-two films and became a global pop culture icon, was a black amalgam of three dinosaur species. Sure, stegosaurus and iguanodon were in the mix, but everyone knows that the most obvious physical traits of Godzilla were down to tyrannosaurus.

Arguably, it's still T. rex's appearance in *Jurassic Park* that created tyrannosaurus's most iconic movie moments. Famously brought back to life using blood from fossilized mosquitoes set in amber, T. rex runs riot in the theme park, attacking visitors and yet sensibly munching on a lawyer. You lose some, you win some. Indeed, the Tyrannosaurus is used for the film's finale, indirectly saving our heroes by totaling the velociraptors, which had been on history's first raptor human hunt in the latter part of the movie.

Given that T. rex is the most famous media dinosaur, so frequently seen on film and in fiction, is its reputation deserved? Was T. rex truly king of the real Jurassic world?

T. Rex in Deep Time

Picture the prehistoric scene. The edmontosaurus was chill. Poised on the bank of a raging river, the duck-billed creature was safe enough from any danger on the opposite bank. And from its sanctuary, this beast was witness to something the trikes realized only when it was far too late. For, across the wild silver-shimmering water, barely more than 40 feet away, a small herd of triceratops was browsing obliviously on the riverbank.

A little way beyond the place where the Trikes browsed was a screen of taller trees, marking the boundary between shoreline and jungle. The dense greenery of the trees was the perfect camouflage for the far more ferocious viridescent beast that lurked behind the screen. Scarcely a sign gave the T. rex away, save its sparkling eyes, the size of baseballs, flicking this way and that in anticipation of its imminent kill.

Tyrannosaurus Attack

All seems calm. River runs, sun sets, birds sing, trikes low. Then, comes the crashing violence. Like a 10-ton truck hurtling through the trees, the scarlet-eyed and green-skinned T. rex must have been a terrifying sight for its prey. Forty feet in length and over 5 tons in weight certainly sounds like a "warlord of the Earth" and "prize fighter of antiquity." The furry fuzz that spiked out of the scales on its back and neck gave the T. rex a punk look. (Sid Vicious, indeed.)

Not forgetting the gargantuan head, ripped frame, muscular legs, and long powerful tail.

The T. rex leads its lunged attack with its gargantuan jaws agape, revealing a mouth full of daggered teeth, gleaming in the setting sunlight. That fierce tyrannosaurus bite tries to puncture the neck frill of one of the trikes, but those neck frills are made mostly of hard bone. The use of the neck frill in dinosaurs is uncertain. Maybe it's for thermoregulation. Or perhaps it's just a defense mechanism. During battles for territory, competing trikes smashed their heads together with their horns, and the neck frill may have been used as a kind of shield, which protected the rest of the trike from harm. It certainly pays off during this T. rex attack. As the first trike hauls its punctured bones into the trees, the T. rex roars in annoyed frustration and hunger, but soon it spies a more likely victim. The smallest of the trike herd sits isolated by the water's edge. Raging river on one side, roaring tyrannosaurus on the other is not where a young trike should be.

A Snarl of Tyrannosauruses

Tyrannosaurus and Trike locked in a deadly stare. The great head lunges once more, the dagger teeth meet flesh. Bones break and blood vomits as the shattered teeth of the Tyrannosaurus rip at its young victim. But wait. There's another twist in this tale. The tyrannosaurus is not alone. From the jungle emerge three more meat-lusting monsters, almost the same size as their leader. The tyrannosauruses were hunting as a pack.

A pack of tyrannosauruses. Hang on. What imaginative collective noun for tyrannosauruses would best suit, dear reader? We all know that dolphins come in schools. Most readers are no doubt aware that one speaks of a murder of crows, a parliament of owls, but perhaps not so many know that it is an exaltation of larks. So, tyrannosauruses? A "bleed," perhaps. How about a "snarl" of tyrannosauruses? Or maybe a "thunder" would be more apt? For that matter, we could have a battalion of ankylosaurs, given their tank-like structure. And maybe a bellow of edmontosaurus, as they seem to have been noisy in their social interactions, given what we know of their nasal cavities and the resonating chambers in some species' crests. (Not forgetting a "menace" of allosaurs, a "dagger" of raptors, and a "mosh" of mosasaurs!)

Back to the pack. Three other red-eyed monsters emerge from the jungle. Branches crack and leaves scatter. The edmontosaurus feels as though its heart is pumping ice around its veins instead of blood. The thunder of tyrannosauruses charge at the riverbank. The three "prize fighters" join with their leader and rip at the flesh of their kill.

Sadly, for the small herd of Trikes, it was a familiar story. On previous occasions they had escaped the hungry jaws of the tyrannosauruses. Maybe they'd managed to gore their predator with the long horn that juts forward from their neck frill. Or perhaps they'd just gotten lucky. But the tyrannosaurus was the nemesis of all triceratops. As Robert Bakker wrote in his 1986 book, *The Dinosaur Heresies,* "The scene has been portrayed in paintings, drawings, and illustrations hundreds of times, but it remains thrilling. Tyrannosaurus, the greatest dinosaur toreador, confronts triceratops, the greatest set of dinosaur horns." According to Bakker, "No matchup between predator and prey has ever been more dramatic." In his view, it was very apt that "those two massive antagonists lived out their co-evolutionary belligerence through the very last days of the very last epoch in the Age of Dinosaurs." Singing the praises of tyrannosaurus, Bakker claimed that "no predatory dinosaur, no predatory land animal of any sort, had more powerful jaws." And Bakker saw two ways of withstanding a tyrannosaurus's attack: "either tank-like armor—the approach taken by ankylosaurus—or the most powerful defensive weapons—the approach taken by triceratops."

A Terminal Cretaceous Arms Race

It was Bakker who described the relationship between T. rex and trike as a "terminal Cretaceous arms race." He cites the close-linked co-evolution of the two species. Bakker wrote:

> Why would triceratops invest in five times as much bone volume in its frill? Well . . . to me the answer is obvious. Because the commonest predator has evolved great, armor-penetrating teeth. The argument goes in the other direction—T. rex evolved swollen, tall, tooth crowns to deal with the unusual protection of the commonest horned herbivore.

T. rex isn't simply a celebrity dinosaur, a movie creation of the twentieth century. No, tyrannosaurus rex was a nightmare made flesh by evolution. The T. rex bite is unique among the more famous dinosaurs. Rather than inflicting long, shallow wounds, the tyrannosaurus jaws would thrust a few crowns deep into bone armor, killing its prey with a single blow.

By now, scientists know quite a lot more about the "Last of the Great Reptiles" than they did when *Jurassic Park* was first released. As *National Geographic* put it in their dinosaur special of September 2020:

> Groundbreaking new science is changing everything we know about how dinosaurs looked, moved, and lived. A fierce but fluffy tyrannosaurus rex. A velociraptor that was more turkey than terror. An armored dinosaur that used camouflage rather than brute force to survive. Over the past few years, a dazzling array of fossil finds, coupled with advances in technology, have dramatically revised our pictures of even the most iconic dinosaur species.

Fierce but Fluffy

A fierce but fluffy tyrannosaurus rex! Indeed, paleontologists can tell how a T. rex looked, how it grew in the way it did, what it ate to get so damned big, and even how it lived and breathed when it was king of its prehistoric domain. The scientific reason we now know so much about T. rex is the sheer weight of fossil evidence. More than fifty skeletons have now been found—that's more than the vast majority of other dinosaurs.

Yet, the most important cultural reason for these advances in T. rex knowledge is this: Scientists are creatures of the culture in which they swim. Paleontologists also read the newspapers, go to the movies, and even play video games. They know the T. rex hype. And that hype feeds back into the science. Some scientists, in fact, are pretty obsessive about tyrannosaurus rex. Good science requires motivation. Great science requires obsession. And obsession is essential to creativity.

So, scientists have thrown everything but the kitchen sink at T. rex. Computational animations to pursue questions about its posture and movement. Software modeling to mull over how T. rex ate. Microscope analysis of its bones to get a grip on how it grew. And CAT scans to think about the function of its sense organs and brain. For example, according to that 2020 *National Geographic* dinosaur special, "The contours of T. rex's braincase show paleontologists that the animal relied heavily on its sense of smell. A 2019 study inferred that T. rex likely had 1.5 times as many genes for odor receptors as humans do, based on the relative size of the brain region that processes scents."

Indeed, paleontologists have recently unleashed a veritable deluge of dinosaur discoveries with such techniques. Scans reveal that major groups of dinosaurs evolved specific cranial air-conditioning systems to make sure their brains didn't overheat. The innovation was this: Large predators, such as T. rex, were able to vent unwanted heat using their large snout sinuses. And, somewhat like a smith using a bellows, T. rex et al. would flex their jaws to push air in and out of their chambers. The innovation made moisture evaporate and wick away heat, like sweat on a sizzling day.

Prize Fighter of Antiquity

As a result of all this technical data, scientists know more about the "king of all kings in the domain of animal life" than they do about some living creatures. So, let's take this alleged "prize fighter of antiquity" and place him in the (blood) red corner of a heavyweight boxing bout. What are T. rex's vital statistics, which could be brought to any prehistoric fight?

An adult tyrannosaurus would check in at 7 or 8 ton in weight and at about 42 feet in length. This is a pretty impressive prize fighter. Who should we put in the blue corner? Well, consider allosaurus, torvosaurus, and their ilk. Sure, these monsters come in at an impressive 33 feet long, and a few ton in weight. But they simply don't rate against the Rex. There's another possible candidate in some of the carcharodontosaurs. Cretaceoun climactic changes meant that some of the carcharodontosaurs, such as giganotosaurus in South America and Africa, grew more than their Jurassic ancestors, getting almost as long as T. rex (yet still an important ton or two lighter).

T. Rex at Close Quarters

T. rex stands alone as the most gargantuan meat-eating monster that lived on land during those days when the dinosaurs ruled. The key to tyrannosaurus power is that head. It's a murder machine. Let's imagine being murdered by the Rex. Okay, we all know the hypothesis in *Jurassic Park* is *very* controversial. Yes, some scientists report extraordinary organic survival in fossils, such as red blood cells in a mastodon or a woolly mammoth that's about a hundred thousand years old. But most scientists have a problem with the idea that DNA and other biomolecules are going to hang around over millions of years on geological timescales. Biomolecules are very decay prone.

Putting scientific objections to one side for the purpose of dramatic spectacle, suppose T. rex is back in town and you, dear reader, are on the menu. It's worth noting, as he approaches you, that his head is about the same length as you, roughly speaking, give the odd foot or two, and you may very well be giving up an odd foot very soon. You might also gulp as you spy around fifty dagger teeth dwelling in that skull, hidden by that mask of death with its sinister smile. And, just as you're worrying about those fifty teeth doing their chomping job on your ass, you also spy smaller "nipping" teeth at the front of the jaw, along with a set of spikes along the jawline. As Rex comes closer, you notice the huge bulging muscles at the back of its head. Their job is simply to open and close those titanic jaws. Those homicidal eyes, the size of baseballs, those satanic horns, that lizard-like crocodilian hide running from that colossal head down to that terrific tail.

But a final twist before your fate is sealed. As you're devoured in practically one crunching gulp, you witness what Lex and Tim Murphy were denied in *Jurassic Park*, because the sci-fi hadn't yet inspired the scientists to find out more about T. rex: Rex has feathers sprouting out between its scales, the longer ones much like the punky quills of a porcupine.

Death Shall Have No Dominion

The Rex's reign lasted around three million years. The actual dominion of this so-called "royal man-eater of the jungle" was the wooded coastal plains and river valleys of the western part of the continent that we now call North America. It reigned over a wide range of ecosystems and preyed upon the same species we find in *Jurassic Park*: the heavily armored ankylosaurus, the duck-billed edmontosaurus, and the dome-headed pachycephalosaurus.

Like many great Americans, Rex was an immigrant. In fact, Rex made a similar journey to the one that immigrant humans made millions of years later to "America." Scientists believe that the human ancestors of the first Americans crossed from Siberia into Alaska. How? The answer is to be found in the last ice age. As sea levels dropped, a new landmass, named Beringia, rose up from beneath the Bering Sea. This new land provided a path from the Russian east coast to Alaska. It's likely that animals were heading for new pastures, and their human hunters simply followed them into new territory, and into the New World. The human journey of migration that had begun in Africa, and divided in Asia, had finally reached the last corner of the Earth, the last continent. This group of humans had found a new home. It had been an incredible journey. They'd survived drought, famine, and an ice age to get there. Before them was an empty continent with lots of roaming buffalo and mammoths.

Millions of years before humans, Rex's migration also seems to have begun in east Asia, probably Mongolia or China. Rex too stomped his way across a Bering land bridge, hunting his way down into the heart of what we now call America. And, as with us humans, when Rex arrived in his new home, he found a continent destined for domination. Rex roved across northern America, migrating up into what we now call Canada, and down into the southern areas now known as Texas and New Mexico.

Rex stood alone, tyrant king of an entire continent. Only something cosmic could knock him off his Cretaceous throne, and it arrived from the sky, in rock form, around 66 million years ago. Yes, T. rex truly was king of the real Jurassic world, until he was blown away by a comet at the peak of his power.

HOW CAN WE KNOW THE BEHAVIOR OF EXTINCT ANIMALS?

"He moves like a bird; lightly, bobbing his head, and you keep still, because you think that maybe his visual acuity's based on movement, like T. rex, he'll lose you if you don't move. But no. Not velociraptor. You stare at him, and he just stares right back."

—Dr. Alan Grant, *Jurassic Park* (1993)

Compelling descriptions like that draw you straight into the dinosaur's world. Well, at least the world as understood by the paleontologist, whose job it is to develop the knowledge that informs such descriptions.

Dr. Grant's explanation gives the impression that scientists have established how dinosaurs moved and also how their vision works. However, science isn't always as final as people often think, and the current understanding of something can get overhauled following updated evidence and the development of new arguments. This is especially true when inferring behavior based on creatures that lived more than 66 million years ago. With so much time separating the dinosaurs and the people researching them, how can scientists know anything about the behaviors of these extinct animals?

Interpreting Behavior

At the very heart of science is the ability to test alternative ideas to build new knowledge, so there are often competing ideas about a particular issue. The scientist's task is to refute these ideas by using and referring to different lines of investigation. Although, some things are easier to investigate than others.

If we want to see how living animals behave, we can go out and look at them, but for extinct animals like non-avian dinosaurs, that's just not possible. Unless, of course, someone manages to orchestrate the dinosaurs' de-extinction and populates an island full of them (which, as it stands, is unlikely in the foreseeable future). There's no doubt that the existence of an island full of dinosaurs would provide the ultimate testing ground for many of the competing ideas put forward about dinosaur behavior. But, then again, the dinosaurs of *Jurassic World* were actually hybrids (crossed with DNA from frogs and who knows what else) so they may not have even displayed attributes and behaviors that were authentic to the original dinosaurs of the Mesozoic.

Even if the dinosaurs were true to their original kin, observation of the few dinosaurs living in a park couldn't provide definitive proof of the behaviors of all known dinosaurs. It's like seeing a few lions and zebras at a zoo and assuming you had a good idea of life on the African Savannah. An example is the common observation that cats hate water, which seems true until you see a Turkish Van cat happily swimming in a lake. It's the same species (i.e., Felis catus), but a different breed showing a unique behavior.

The point is that behavior is complex, and we really can't assume that what fits for one, will apply to the other. Even individuals of the same species within the same family can show different specific behaviors based on things like age, sex, and season, so how can paleontologists be confident of anything they say about dinosaur behavior?

Well, ultimately it comes down to the reliability of their investigations and the available support for their claims (i.e., the evidence).

Lines of Investigation

Paleontologists have to interpret various types of evidence to develop possible narratives about what dinosaurs were like and how they lived. This evidence comes from many fields that draw upon science, technology, engineering, and math in different ways. For example, Ichnology is the scientific study of tracks (i.e., footprints and trails) and traces such as nests, eggs, burrows, borings, bite marks, and coprolites. These are useful for showing what animals got up to while living, rather than what happened at death, which is when body fossils are laid down.

A useful investigative technique is biomechanical modeling, which looks at how limbs could have moved and functioned by making physical or virtual models of the arrangements of limbs. These models can provide insight into a creature's potential range and speed of movements, as well as the forces they could have applied in doing so. Of course, these models are strongly affected by the assumptions used to create them, such as the position of soft tissue, the strength of muscles and bones, etc. However, by using the same models on a living creature, it's possible to get an idea of how well the modeling technique compares to reality.

On the whole, to get an idea of what's actually possible in nature, paleontologists make extensive use of analogy with living creatures. This might involve looking at where muscles and bones are connected in extant (i.e., not extinct) relatives of dinosaurs or even in animals that have similar characteristics such as being heavy, fast, carnivorous, etc. This can help to identify any anatomical specializations that might have also been observed in dinosaurs. For example, carnivores tend to have sharp teeth and claws, heavy animals have load-bearing limb arrangements, and animals suited to running (called cursorial) favor relatively slender leg bones.

Scientists can also investigate a bone's microscopic anatomy (or microanatomy) by using histology, which studies the cells and tissues through a microscope to examine how their structure relates to their function. This approach has been used to illuminate details like how fast dinosaurs grew and to what age they reached, among other things. There's even a clever way to infer characteristics that aren't directly accessible from fossilized remains, and it can be applied to both physical and behavioral traits. It's called the extant phylogenetic bracket (EPB).

The Extant Phylogenetic Bracket

The extant phylogenetic bracket is an absolutely genius technique introduced in 1995 by American paleontologist Lawrence "Larry" Witmer. The basic idea of it is this: When you look at those diagrams with branching lines showing how different species are related, you're looking at a cladogram. Every node (the place where a line splits) represents the branching of a new group (or clade) of organisms with a shared characteristic. For example, they all have a backbone, or they all lay eggs. The line preceding that split represents the most recent or last common ancestor of all life forms directly descended from that point on the cladogram.

If a particular characteristic is present in a group that branched off earlier in history (more basal) than an organism, and the characteristic was still around in a group that branched off later in history (more derived) than the organism, it can be inferred that this trait also existed in the organism itself. The basal and derived groups are said to bracket the organism. This means that if we can find extant descendants of groups that bracketed the dinosaurs, and both groups share a trait or characteristic, we can infer that this characteristic was also present in dinosaurs, even though the dinosaurs are extinct.

The closest extant groups that bracket the dinosaurs are birds (more derived) and crocodiles (more basal). The last common ancestor of these groups (and all its descendants) is known as an archosaur. Using the EPB, paleontologists can look at characteristics common to birds and crocodiles to infer the presence of these characteristics in other archosaur descendants, like dinosaurs. What's so powerful is that these characteristics are not limited to physical features but can also indicate the presence of particular genes, as well as potential behavioral traits.

Armed with tools like the EPB and the other lines of investigation mentioned above, scientists can carefully reconstruct what may have happened in the past using evidence rather than speculation. Although each investigative technique may have its limitations, a fuller picture can be created when the evidence-based insights they each reveal are carefully cross-referenced and added up.

Now that we know a bit more about how scientists can build up a degree of certainty about the past, let's see how reliable Grant's narrative would have really been regarding current ideas on potential dinosaur behavior.

Walking the Walk

The most primitive form of walking uses a sprawling stance with limbs jutting sideways from the body and bellies near to the ground. This is how lizards walk, and it used to be thought that dinosaurs moved that way, too. However, by scrutinizing the range of movements that their bones could allow and by observing fossilized trackways, the evidence appeared more consistent that dinosaurs stood and walked with an erect stance, where their legs were mostly held directly beneath them and tails held in the air. This was the image of dinosaurs that *Jurassic Park* portrayed and helped to popularize in the public's imagination.

Genuine remnants of dinosaurs in motion are preserved in their footprints. By measuring the dimensions of a single footprint, scientists can estimate the potential size of a dinosaur and what major group it may have come from. For example, sauropod, theropod, ornithopod, or ceratopsian. Taken in isolation, though, the tracks can't identify a species. To do this, scientists look to evidence from dinosaur body fossils found locally to see which of them best correlates with the size and type of impression.

A series of two or more footprints is more informative. These are known as trackways and can reveal whether an animal walked on two legs (bipedal) or four legs (quadrupedal). They can also indicate whether dinosaurs were walking, running, or even wading through water at the time they laid the tracks. By measuring details like the length of a single step (a pace) and the distance between placements of the same foot (a stride), the dinosaur's approximate hip height, size, and speed can be determined using formulas. By combining these insights with other details of the ground and surroundings, scientists can determine what the animals appear to be doing, and what other things were in their environment at the time.

Trackways are very useful, but what about identifying the dinosaurs' mannerisms, such as Dr. Grant's description of a velociraptor moving "like a bird, lightly bobbing its head"? Well, the majority of birds don't actually bob their heads, and it's believed that the ones that do, do it either to improve vision or to maintain balance.

A popularly cited study by Dr. Barrie J. Frost in 1978 looked at the apparent head-bobbing movement of pigeons. In the study, he found that the forward- and back-bobbing movement turned out to be an illusion. In fact, when walking, they tended to keep their heads as stationary as possible to reduce any motion blur in their eyes while walking and viewing their surroundings. Regarding balance, birds have shorter tails and heavy flight muscles in their chests, so most of their weight is positioned forward rather than above their legs. As a result, on each step, the head moves as a counterbalance to stop them from toppling over. According to paleontologist David Hone, "I would not expect many bird-like theropods to bob, as they had long bony tails and far smaller chest muscles, meaning their balance was much better than a bird's." He also added that if they did bob, it would only be a little.

Built for Speed?

A huge part of the Jurassic series is the visualization of carnivorous dinosaurs chasing down herbivores and humans. While trackways can allow estimations of the speeds of the creatures who made them, those calculations are limited to the specific tracks. Scientists have other means to estimate a dinosaur's speed, though.

By looking at living animals, scientists can see how the form of an animal's features relates to their function. This is known as functional morphology, often encapsulated in the adage: "Form follows function." For example, the relatively long shin bones of cheetahs and ostriches favor their cursorial (suited to running at speed) nature. Velociraptor bones have a similar form, indicating that they too were cursorial creatures; hence, the name velociraptor, which means swift plunderer. Additionally, their legs were quite muscular, along with a stiff tail which helped them to maneuver. These features were also present on all of their dromaeosaur relatives, such as the larger deinonychus or the even bigger achillobator. It's estimated that velociraptor may have achieved speeds of up to 40 mph.

Another method of estimating dinosaurs' speeds is to look at the biomechanics of their skeletons to calculate how their muscles would have pulled on their limbs and with what force. The muscle arrangements can be estimated by analog with living creatures but also via bumps and marks on the bones that can reveal where muscles were attached.

The caudofemoralis links the tail to the femur and is the main muscle used by dinosaurs in running. If the muscle is attached near the end of the bone, it allows faster movement of the limb, but at the expense of requiring more energy and tiring the animal quicker. This was the case with T. rex. Conversely, if the muscle is attached farther down the bone, it would be easier to move the load, which is great for endurance but makes limb movement slower. This was the case for hadrosaurs.

Dinosaur Chase Scenes

The specific details of a T. rex in the chase were investigated in 2014 by University of Alberta paleontologists Scott Persons and Phil Currie, who revealed that in a race between a T. rex and a hadrosaur, the T. rex could catch up easily but over longer distances the T. rex would tire, enabling the hadrosaur to escape. As such, tyrannosaur specialist Scott Persons suggested that T. rex would have been better off ambushing such prey, which is something we see the T. rex do frequently in the movies, whether to a gallimimus herd, a pack of velociraptors, or a menacing carnotaurus.

Over the years, many estimations have been made about the running speed of T. rex and the views have changed considerably, ranging from 11 mph to 45 mph. In 2017, a UK-based team of researchers set out to combine two separate biomechanical techniques to establish a more accurate estimate. They found that a T. rex's long limb bones would have undergone too much stress at high running speeds, limiting the adult T. rex to a quick walk instead. They indicate that it had a top speed of about 17 mph, which is like doing a 100-meter sprint in 13 seconds. Although, if the T. rex were a juvenile, it would be much lighter and could run faster than its older counterparts before skeletal stress became a limiting factor.

It turns out that T. rex was quite the ballerina, though, based on a study published in 2019, which demonstrated that T. rex was very good at turning. According to one of the researchers, Eric Snively, this ability suggests that T. rex could "successfully attack smaller, younger and/or more dangerous prey than other carnivorous dinosaurs would bother to tackle." He also remarked that the juveniles were much scarier.

So, what do you do if you ever get cornered by a Cretaceous carnivore? Well, it depends on the animal's size. If you were quite fast, you could try running, but you'd only outrun the biggest ones, while the smaller or more agile younger dinosaurs would still easily catch you. So, that infamous Jeep-chasing scene would have probably involved an older T. rex walking quickly for a bit and then stopping, giving an aggravated rumble, and maybe even sticking up its two fingers in consternation before turning and walking back into the forest. However, if it was a juvenile, the chase would have lasted a bit longer due to its added speed and agility.

Now You See Me

Running could be an option when faced with a T. rex, but didn't Grant mention something about visual acuity? Well, visual acuity refers to the clarity of eyesight (i.e., the smallest letters that could be read on an optician's Snellen chart). But the infamous optician's test is static with typically simple color and contrasts. It's good as a comparative tool but isn't the best approximation for the more natural circumstances of vision such as differentiating objects that have low contrast between them, tracking moving objects, or viewing things in low light conditions. So, how could we assume to know about a dinosaur's vision?

It's now known that many dinosaurs had relatively good eyesight, which includes color vision, as identified by extant phylogenetic bracketing. But, it's even possible to tell whether their eyes were adapted for day, night, or twilight viewing. This is done by looking at a feature called the sclerotic ring, which is made up of a ring of bones that sit in the orbits (eye sockets) of most vertebrates, excluding crocodilians and mammals. The bigger the animal, the bigger the ring, with the diameter of the central hole indicating how much the animal's pupils can dilate (i.e., how much light the eyes can gather). Nocturnal animals have a large ring, allowing more light into the eye, while diurnal (daytime) creatures have a smaller ring, which permits less light but also provides a sharper image. Crepuscular (i.e., twilight) animals fall somewhere between those extremes.

Back in 2006, US researcher Kent Stevens investigated the visual acuity of seven theropods including T. rex and velociraptor. He sculpted reconstructions of their heads to map out their binocular field of view (BFoV). Binocular vision is a key feature of predators, providing better depth perception, better ability to recognize objects, and also providing an advantage for nocturnal vision. Stevens found that some dinosaurs such as allosaurus had a BFoV covering only 20 degrees, while tyrannosaurus and velociraptor had a better BFoV at 45 to 60 degrees. This meant that they had a bigger region in which the view from each eye overlaps; a region that would have provided better depth perception, clearer images, and an increased ability to distinguish objects from their backgrounds. So, both T. rex and Velociraptor had good eyesight.

Now, if you were caught in a predicament on Isla Nublar and lucky enough to be apprehended by a slow and maybe aging dinosaur with poor vision, it's worth noting that T. rex and velociraptor both had a great sense of smell. Particularly, T. rex had huge olfactory bulbs, which would have made it excellent at sniffing out prey. This was identified in a 2008 study led by Darla Zelenitsky in Canada, who noted that "Large olfactory bulbs are found in living birds and mammals that rely heavily on smell to find meat, in animals that are active at night, and in those animals that patrol large areas."

The size of the olfactory bulbs was established by scanning and measuring dinosaur skulls and analyzing the internal impressions left by the presence of the animal's brain (and associated olfactory bulb) in life. This is also supported by the size of the space left by nerves that would have conveyed this olfactory information to the brain. Also, in 2019, researchers in Ireland published a study that made use of information about olfactory receptor genes in living dinosaur relatives. They projected this information back to the dinosaurs using the EPB to ascertain which of these genes may have been present within particular dinosaurs, including T. rex.

The results supported previous indications that tyrannosaurs had an extremely keen sense of smell.

Follow the Evidence

So, there it is. Well, a tiny part of it, at least. Some of the many methods used to establish an evidence base for describing some of the behaviors of dinosaurs. Every single insight acts to bolster or refute a previous idea and as time progresses new technologies and techniques revolutionize the scope of what can be known.

Scientists have interpreted fossilized trackways, scrutinized fossils using histological techniques, and drawn comparisons to extant life forms. They've introduced the extant phylogenetic bracket, created virtual biomechanical models, and integrated the use of genetic information. As a result of their careful and considered contemplations, we now have a well-informed idea of how dinosaurs may have behaved in the far reaches of Earth's history.

We know about their mannerisms, how they moved around, and at what speed, as well as how they literally viewed their world. The assumptions are not always spot-on, but as with all science, the conclusions should be regarded as the current take on the knowledge accumulated thus far. This knowledge has inspired many narratives of a remarkable past and Michael Crichton's Jurassic series is just one of them. But the story is far from complete, and rest assured that the evidence scientists uncover will continue to astound and inspire as time rolls on.

HOW DID DINOSAURS ACQUIRE FEATHERS?

Featherless Fiasco

Back in 2015, *Jurassic World* got a fair amount of flack about feathers. Dinosaur dweebs and paleontologists the world over were mad at the moviemakers. The much-anticipated return of the 20-year-old franchise was due to finally arrive. But, as it happened, without feathers. Without the software tweak needed to render the franchise's ferocious film stars feathered and bleeding-edge, scientifically speaking.

In particular, people were expecting that the production of *Jurassic World* would update dinosaur species such as the film's velociraptors and maybe even T. rex itself. Feathers, or feather-like features, would be on the menu, they thought. But, when early production photos were released, many commentators were gutted to find the franchise's "raptor" pack as naked as in 1993's *Jurassic Park*.

It's Sci-Fi, Not Science

And yet the franchise seemed to speculate its way out of this scientific controversy. One of the traditions of science fiction like *Jurassic World* is the spirit of the so-called "what if?" question. There are great similarities between science fiction and science. Science fiction is an imaginative device for doing a kind of theoretical science, the exploration of imagined worlds. Scientists build models of hypothetical worlds, and then test their theories. Einstein was famous for this. His thought experiments, *Gedankenexperiment*, led to his theory of special relativity, for example. The science fiction moviemaker also explores hypothetical worlds. But with more scope. Scientists are meant to stay within bounded laws. The likes of Steven Spielberg and Colin Trevorrow have no such boundaries. However, we can see that a spirit of "what if?" is common to both science and science fiction.

The *Jurassic World* production created an in-universe argument for its decision about dinosaur appearance and threaded that argument throughout the movie. Perhaps the best example of this is when the Park geneticist Dr. Henry Wu is confronted by Park owner Simon Masrani. Masrani's charge is that Wu designed and engineered the absurdly dangerous indominus rex. But Masrani, in the interests of corporate capitalism, had demanded new and more terrifying dinosaurs for the Park. (Recall once more the words of Claire Dearing: "Stegosaurus is like an elephant now to kids.") They wanted it created quickly, and secretly. It was a blueprint and a schedule that produced an animal with a violent temper and the dinosaur equivalent of social dysfunction. Masrani tells Wu that genetic tinkering with the indominus was a mistake.

But it's Dr. Wu's answer that is revealing here. Wu simply points out that his work on the indominus is no different to the rest of his career to date. Indeed, Wu says, none of the dinosaurs in Jurassic World, or its previous incarnation as Jurassic Park, are pure dinosaur DNA. None of the Park's creatures are what they were 66, or more, million years ago. In other words, Wu implies, he created what he was asked to make: dinosaurs that look exactly as the public would expect. Scary, yes. Ferocious, for sure. Scaly, preferably, but not necessarily feathered.

The Chinese Revolution

Why did the planet expect the dinosaurs of *Jurassic World* to be far more feathered than their franchise counterparts 20 years earlier? In a word, China. As we mentioned in the introduction to this book, fossil finds in China since *Jurassic Park* have since shown that many dinosaurs were feathered. Not only that, but some dinosaur species managed to survive the great extinctions and are the ancestors of our modern birds. These recently found Chinese fossils of feathered dinosaurs are so well preserved that paleontologists can fathom the feathers' color and where they were found on the dinosaurs' bodies. As a result, theories have emerged about the use of feathers for display, insulation, and in some cases maybe, flight. Little wonder *Jurassic World* came with a feather forecast.

Since the 1990s, Chinese workers, such as farmers, scholars, and fossil dealers, have brought to Beijing thousands of fossils from Liaoning Province in northeastern China. These fossil finds have revolutionized our understanding of how dinosaurs looked and behaved. Many fossils have traces of feathers, which show that plumage first evolved before dinosaurs flew. A number of fossils also suggest, rather dramatically, that dinosaurs other than birds' closest ancestors also toyed with flight.

Dinosaur Classes

There are three main groups of dinosaurs. The long-necked beasts like brontosaurus are known as the sauropod dinosaurs. There are the plant eaters, which are known as the ornithischian dinosaurs, creatures like triceratops and stegosaurus. And then there's the theropod dinosaurs, the meat eaters.

This third group, the theropods, are the dinosaur group most associated with feathers. Birds come from theropod dinosaurs. They evolved from theropods in a similar way to how humans evolved from apes. So, birds *are* theropod dinosaurs. Now, we might ask what it was that distinguished theropod dinosaurs to evolve and carry feathers. Those fossil finds from China over the last couple of decades suggest that feathers go way back to the earliest dinosaurs.

The fossil evidence suggests that one group of the theropods gradually got smaller, changed their skeleton through genetic mutation, evolved feathers, mutated *those* feathers, and eventually evolved into birds. So, some theropods are meat monsters like T. rex, but there was great diversity in these theropods. Many had crests and sails on their backs, and birds emerged from such creatures over fifty million years.

Dinosaur Feathers

What exactly do we mean by dinosaur "feathers"? Feathers are incredible integumentary structures. An integument means an outer covering, such as skin or shell, and an integumentary system can include the skin, along with its appendages, which act to protect the body. Integumentary systems can include hair, scales, feathers, hooves, and nails. Scientists know from the study of the development of feathers in chick embryos that, as the embryo is developing, a little thickened region is formed in the outermost part of the skin. This thickened region then starts to project inward. It forms a follicle, and the cells lining that follicle, which are from the dermis or the lower part of the skin, die off and start to form the interior part of the feather.

At the same time, some cells project outward from the skin to form a hollow shaft. If you have ever held a feather, you'll know that this is basically the fundamental structure of a feather. It's a hollow tube. Admittedly, some modern birds have more complex feathers. They're actually the most complex structures derived from the skin in vertebrates. So, a typical feather such as a flight feather in a crow or a jackdaw actually has several levels of branching. It has a central shaft or tube, what we know as the quill part of the feather, which branches into smaller structures called barbs. Those barbs have little branches called barbules and these barbules are in turn differentiated at their tips into little hooklets that can zip together to form a nice closed vein.

What about feathers on dinosaurs? Well, when scientists look at the dinosaur tree, at the theropods and the ornithischian dinosaurs, they find evidence of feathers at different developmental stages, different evolutionary stages. In short, they find a spectrum of fossil feathers, from very complex feathers that are anatomically modern, looking identical to today's birds, all the way down to simple filaments that are basically hairs.

Chinese Dinosaur Feathers

The Chinese feathered dinosaurs are the most crucial fossils to have been found for many generations, certainly the most important since *Jurassic Park*. They're vital because their preservation is first class, and because there are so many different dinosaur species found with feathers. The Chinese fossil finds span the dinosaur family tree. And when scientists map out which species on the tree have feathers and study how those feathers change, it tells the tale not only of feather evolution but also of bird evolution—that is, how some dinosaurs turned into birds.

It seems that any theropod dinosaur found in China has feathers of some kind. Even the tyrannosaurs, such as yutyrannus, have coats of very simple hair-like feathers. There are also lots of other small theropods with simple feathers. The same goes for plant-eating dinosaurs. So, scientists believe that feathers go all the way back to the very beginning of dinosaur evolution. And it's probably true that all dinosaurs had some type of feather.

But one particular group of theropods, the maniraptoran theropods, began to lengthen those feathers into barbs and barbules and began to line up feathers on their arms. Wings formed. These dinosaurs got smaller and it was this group that eventually evolved into birds. Scientists now also think that hairs, scales, and feathers all ultimately evolved from the same primitive structure.

Feather Colors: Ginger Dinosaur?

Paleontologists can even tell what color some feathers were. If you crack open the feathers of modern birds, and place them under a powerful electron microscope, you will spy tiny, microscopic structures. They can appear to be little ball-like structures, or tiny sausage structures. Both structures are granules of the pigment melanin. And melanin is just one of the many pigments modern birds have in their feathers that give them color. What's really remarkable is that, since *Jurassic Park*, dinosaur scientists have realized that evidence of melanin pigment can survive in fossil feathers.

So, scholars have studied the melanin granules, or melanosomes, and have noted that different shaped melanosomes generate different colors. The sausage-shaped melanosomes produce blacks and browns, and the ball-like melanosomes produce reddish, foxy, or ginger colors.

For example, consider sinosauropteryx. Meaning "Chinese reptilian wing," sinosauropteryx was first reported in 1996. It was the first dinosaur outside of birds and their immediate relatives to be found with evidence of feathers. The creature was covered in a coat of simple filament-like feathers. But coloration has also been preserved in melanosomes in some of its feathers, making sinosauropteryx the first dinosaur where coloration has been found.

And it seems that sinosauropteryx was a ginger dinosaur. The dinosaur's tail was striped like a barber's pole, with repeated and regular stripes of ginger and white. The remainder of its body was covered in a sort of ginger color over the back, and maybe pale on the belly, with various patterns over the face. Why would sinosauropteryx have a stripy tail? Camouflage like that of a tiger or a zebra would seem unlikely, as it's only the tail that's patterned, so it's far more likely to have been for display. Sinosauropteryx is likely to have hopped around, waving its tail, and declaring the dinosaur equivalent of "check out my ass."

Feather Conclusions

Scientists now believe that feather evolution and the evolution of wings happened more like a mosaic. This mosaic is put together from fossil evidence which appears to show that the hind limbs, fore limbs, and tail all evolve almost independently. And these changes toward a bird-like appearance are happening in piecemeal spurts in different dinosaurs.

The process of dinosaur into bird would have begun over 200 million years ago with the origin of dinosaurs. Then, at 150 million years came the famous archaeopteryx, a genus of bird-like dinosaur that is transitional between non-avian feathered dinosaurs and modern birds. Archaeopteryx had proper muscular wing beats and powered flight and, most people would say, was the first bird. The Chinese fossil finds have shown us that there is now a good record of feathers and birds right through to 66 million years ago. With the extinction of the dinosaurs came the extinction of a lot of these early bird types. But those that survived led to the massive explosion of modern birds and 10,000 species of birds today. Dinosaurs live on.

DID DINOSAURS OF A FEATHER FLOCK TOGETHER?

Alan Grant: "Look at the wheeling, uniform direction changes. Just like a flock of birds evading a predator."

Tim Murphy: "They're uh . . . they're flocking this way."

—*Jurassic Park* (1993)

Jurassic Park has loads of scenes where groups of dinosaurs are living together happily. Whether it's sharing the same habitat for feeding (parasaurolophus), traveling in groups (brachiosaurs), hunting together (velociraptors in nearly every movie), or running from a common danger, such as the gallimimuses fleeing from a T. rex.

When it comes to social behavior, some animals are more socially inclined (known as gregarious), while others are more solitary, but it's not a simple case of saying that a particular species always behaves one way or another. Even within a species the social behavior of the individuals can vary depending on several factors.

Seeing as birds are a surviving group of dinosaurs and that both groups have a range of physical features in common (such as feathers, egg laying, and bipedalism), would dinosaurs have also flocked together like many birds do or would they just come across as lizard-like loners?

Social Animals

Most animals are social to some degree. Solitary animals are the least social, only really interacting for breeding reasons. They'd much rather roam their own territory and have no other contenders for their food, shelter, and breeding opportunities. This solitary behavior is more common in carnivores, but a few herbivores live this way, too, such as koalas and pandas.

Whether or not an animal is social has a lot to do with whether it provides a survival advantage or not. For example, even the most solitary animals have to eventually locate or attract a mate in order to pass on their genes, and they'll adopt whatever behaviors make that act more probable. The same applies to improving food intake and staying safe in general. So, if it helps a creature to survive or in the very least is not detrimental to its survival, then social behavior becomes part of its lifestyle.

Other than breeding-related behaviors (i.e., mating and caring for offspring), animals may also come together to forage, which increases their chances of locating or catching food. For predators, this means having more teeth and claws to bring down prey. For herbivores, it means having more individuals on the search for food.

Being around others of the same species also offers added protection because there are more eyes on the lookout and more teeth and hooves to strike out against predators. Being among other prey also decreases the likelihood of any particular individual being eaten. Consider the chances of being caught as one of a group of 10, compared to one of a group of 100 animals. This is known as the dilution effect. There's also another consequence, called the confusion effect, in which more targets in a group give predators too much to focus on, overwhelming their attention and leading to a decrease in their attack success rate.

In all of these cases, being social improves the animals' chances of surviving and passing on their genes. Their resulting offspring learn the related social behaviors as part of their upbringing and subsequently adopt them into their lifestyle. It isn't one size fits all, though, and the animals rely on sociality in different ways, with their range of needs being as varied as their range of body forms. And whether they're highly dependent on social interaction or less so, it's clear that many animals gather in assemblages at some point in their lives and this is partly reflected in the sheer number of names humans have given to those groupings.

For example, we're more than familiar with describing groups of domesticated and wild animals as herds. Herds are usually herbivores as well as prey for other animals, including humans. On the other hand, the animals that do the hunting (i.e., the predators) are often described as packs when in groups. These group descriptions are called collective nouns and they can even be specific to particular animals (e.g., a "murder" of crows or a "terror" of tyrannosaurs). Some even get different names depending on what they're doing, with vultures being the most notable. A "kettle" of vultures circle overhead, then land on a tree to become a "committee," "venue," or "volt." If you see them feeding on a carcass, though, they are now a "wake" of vultures.

Historically, these specific labels were linked to hunting and referred to as terms of venery, an old European term for hunting. However, the more general collective nouns like herd, pack, and flock were still used to describe an assemblage of conspecific animals (i.e., of the same species). So, clearly, social interaction is a widespread feature of many modern animals, but what evidence is there for dinosaurs flocking or gathering in groups?

Dinosaur Deathbeds

One common source of evidence used to identify gregarious behavior is bone beds, which are large collections of animal bones found in a particular stratum (i.e., geological layer). These assemblages provide a record of the final resting place of bones that belonged to one or more species of animals that lived around the same time. One example is the Pipestone Creek bone bed in Canada, which is a mega-dense dinosaur graveyard containing up to 200 bones per square meter. It's estimated that two-thirds of its volume is comprised of bones.

Some of the bones belonged to carnivores such as tyrannosaurs, troodons, and a close relative of velociraptor called boreonykus. However, the majority of the bones belonged to both adults and juveniles of a ceratopsian called pachyrhinosaurus lakusti, making it the monodominant group on the site. As monodominant bone beds feature mostly a single taxon buried at the same time, it's generally interpreted that the animals were living together near their time of death, rather than them being positioned together after death by some other potential cause. In *Jurassic World*, after the indominus rex has slaughtered a herd of apatosauruses, their remains (if somehow preserved) would form a monodominant bone bed for any paleontologists exploring Isla Nublar from the distant future.

However, unlike Owen Grady and Claire Dearing stumbling across the death scene, it's often not known for sure how an assemblage of animals came to be together in the first place, or how they died. Even so, the current thinking about the Pipestone Creek bone bed is that it represents a mass-death assemblage caused when a flash flood engulfed the pachyrhinosaur group, flushing their bodies downstream where they got deposited and buried to form the Late Cretaceous pileup.

In another Late Cretaceous bone bed, this time in Mongolia, scientists from the University of Alberta explored an assemblage of remains from a bird-like theropod called avimimus. This time there were numerous remains of adults and subadults but a scarcity of juveniles. This led the researchers to remark that this "may be evidence of a transient age-segregated herd or 'flock,' but the behavior responsible for this assemblage is unclear." Despite the uncertainty surrounding the reason for the assemblage, the find is described as "the first evidence of gregarious behavior in oviraptorosaurs."

So, there you have it, proof that dinosaurs mingled in groups and made flocks! Well actually, that's not entirely true . . . or even close, as we're reminded by David Hone in an online interview with paleontologist Dinosaur George. "It's extraordinarily dangerous to extrapolate across a very big group and say something is universal when it comes to behavior." Dr. Hone later adds that even when we find remains of big groups of dinosaurs, "it's quite dangerous to necessarily say that they actually lived in groups. Lots of animals shift their behavior as they grow, or over the years."

Although there's evidence that some dinosaurs inhabited a same space and time, it doesn't necessarily mean that this was the norm for them or even that it definitely applies to all members of their species. However, seeing a behavior in one animal is at least enough to rule out the possibility that no dinosaurs ever exhibited that behavior. It also allows us to confirm that a particular behavior had already existed back in the Mesozoic. For example, in 2019, scientists revealed 15 nests in the Gobi Desert, Mongolia, in what has been described as "the first clear example of group nesting activities in dinosaurs." Nesting colonies like this are seen in birds and can provide added protection against predators. It's believed that the theropods who laid these eggs likely did it for the same reasons. Finds like these can show us that at least some of the social behaviors we see in modern birds may have already developed within their dinosaur predecessors many millions of years ago.

It should be noted though that in modern birds, some nest together but don't forage together, while others forage together but don't nest together. Basically, social behavior in one circumstance doesn't directly imply sociality in others. So, these bone beds and group nests provide evidence that dinosaurs could be around each other, but don't provide evidence of whether they moved around together in herds. For that, we need to look elsewhere.

They're Moving in Herds

The most common evidence for herding behavior comes from trackways. For example, more than a century ago, dinosaur footprints were recognized at the Paluxy River in Texas, with the first major trackway being excavated there in the 1940s by paleontologist Roland T. Bird from the American Museum of Natural History. These fossilized footprints were identified as belonging to early cretaceous sauropods, and the fact that they ran parallel and in the same direction indicated the animals were all moving together with a common destination or purpose. There are even tracks belonging to meat-eating dinosaurs, one of which was likely stalking the sauropods, judging by the proximity and path of the dinosaurs' footprints.

Hundreds of dinosaur trackways have been found and studied all across the globe, indicating that traveling together was a common behavior in many dinosaur groups. In fact, just before the release of *Jurassic Park* in 1993, a sauropod trackway was uncovered in Portugal, revealing a herd of seven juveniles walking in the same direction, apparently accompanied by three larger individuals. So, scientists have known for a while that at least some dinosaurs traveled in herds, although they can't often tell exactly which species were involved or the reason why the animals were moving together. However, the tracks do sometimes contain information about where and when they were laid, as well as indicating how the dinosaurs may have been interacting at the time.

A trackway discovered in Kenilworth Coal Mine, Utah, revealed signs of dinosaurs walking in a swamp. Coal usually indicates a swampy environment, where ancient decaying plants formed into peat, which became buried (and compressed), eventually turning into coal. The site had theropod and ceratopsian tracks but also an abundance of hadrosaur tracks, indicating a potential herd. It was even determined that most were headed south, with speculation that they may have been making a seasonal migration. Judging by the tracks, the animals were walking leisurely, while also stopping to feed on trees, evidenced by petrified tree root systems that are visible among some of the fossilized hadrosaur footprints.

In 2014, a team reported on a track site found in Denali National Park and Preserve, Alaska. Many of the tracks preserved skin impressions and also had dimensions that matched trace fossils called hadrosauropodus (i.e., tracks related to hadrosaurids). Further analysis also revealed that the hadrosaurs that made them represented a wide range of ages (from very young to adult) living together as a large social group. This hadrosaurid multi-generational group behavior hadn't been seen before in bone beds or track assemblages. For reference, the distinctive Parasaurolophus, which features in much of the Jurassic series, is a well-known member of the hadrosaurid family. Although, as warned before, we can't know for certain whether they behaved this way too.

Now, if you know the scene quoted at the start of this chapter, you might remember how the characters see a herd of gallimimuses running in the distance before changing direction and heading toward them. The original scene in the book actually involved hadrosaurs, although whether any variety of dinosaurs stampeded from predators in this way is still pretty much unknown from the fossil record. In fact, the only known trackway interpreted as evidence of a dinosaur stampede was discovered in the 1960s at Lark Quarry, Australia. Its interpretation as a stampede has been called into question by some but nonetheless this important trackway had a conservation site built around it in 2002. The site is called Dinosaur Stampede National Monument.

Despite the paucity of fossilized evidence, though, a stampede is just a bunch of animals running together from a perceived threat, so it's very likely that some dinosaur herds did stampede, as is seen in many extant creatures including cattle, horses, elephants, humans, ostriches, and even ducks. Maybe future research can link trackway evidence to observations on how living species interact when flocking and swarming, but as yet this hasn't been seen.

There were clearly many different types of dinosaur groups herding and moving around together. There were groups of juveniles walking together with the adults separate, and other whole family units or at least multiple generations moving together. Other examples of dinosaur groups on the move provide evidence of foraging, migration, and even signs of animals being followed by predators. But one more important question still remains about dinosaur social behavior. Did any of those predator species actually move or hunt in packs?

Prowling in Packs

Just to reiterate from the start of the chapter, both the hunted and the hunters can reap benefits from gregarious behavior, and if it helps them survive as a species, it's more likely to be something they do. Although, with the attackers it usually involves a certain extra level of cooperation, rather than just foraging and traveling together for safety in numbers. These predators need to work as a team in hunting and catching food, whether bringing down larger prey like lionesses attacking a hippo or elephant or snaring smaller prey like dolphins creating mud nets to herd and catch fish.

You might recall that scene with the annoying kid in the first *Jurassic Park* movie, when Alan Grant says, "Velociraptor's a pack hunter, you see, he uses coordinated attack patterns, and he is out in force today." This is a behavior shown by velociraptors in all of the movies and was thought to be the case based on fossilized remains of their relative deinonychus. Deinonychus remains were frequently found in the vicinity of a large herbivore called tenontosaurus, which at seven meters long was deemed too big for a single deinonychus to bring down. So, the thinking was that they must have teamed up as a pack to attack the tenontosaurus and, seeing as velociraptors are close relatives of Deinonychus, it was implied that they potentially displayed the same behavior. However, a paper published in 2020 turned that view on its head.

The US researchers noted that the assumption of deinonychus hunting as a group was based on a dog-like social hunting structure, which they saw as strange since modern archosaurs aren't commonly seen to use such "highly coordinated hunting strategies." They put across another idea that maybe deinonychus behaved more like reptilian carnivores such as Komodo dragons. The difference being that Komodo dragons are asocial animals that rear their young in different ways than social animals. Basically, in the asocial animals, the young have to feed themselves and so have a different diet than their parents, whereas in social animals that feed their young, the adults and juveniles generally have the same diet.

Remarkably, these dietary differences are recorded in their teeth, visible as different levels of a specific carbon isotope. In the asocial animals they tested, the younger animals (with small teeth) had more of the isotope and the older animals (with large teeth) had less. When they looked at fossilized deinonychus specimens, they found the same relationship indicating that deinonychus was likely an asocial animal that didn't feed its young. Additionally, the carbon isotopes in the older animals even matched those present in their tenontosaurus food source, while the younger ones didn't. The researchers proposed that, instead, they likely ate a diet based on smaller species from a different part of the food web. Due to their apparent asocial nature, the researchers conclude that deinonychus "was not a complex social hunter by modern mammalian standards."

There's another unexpected twist in the *Jurassic Park* pack hunter story, though. It appears that the other prominent predator of the Jurassic series may not have led such a solitary existence after all. In 2014, Canadian researchers reported on the first-ever tyrannosaur trackway. The tracks revealed three creatures moving next to each other in the same direction. They were adults in their late twenties judging by the size of their footprints. If this was the norm for tyrannosaurs, it may be that they were socially inclined. So, it appears that these tyrannosaurs of a feather did actually flock together, for that moment in time at least.

Let's Get the Flock Out of Here!

Social gatherings are a part of many animals' lives at some point, but on the whole, some animals seek others of their species, while others tend to avoid or be more hostile to those of their own species. It all depends on how well it helps the species to survive. Some stick together for defense, others for attack, but overall sociality brings benefits toward finding food, breeding, and rearing their young.

The fossil record holds many clues to dinosaur assemblages, whether in bone beds or trackways. Although an assemblage doesn't necessarily imply gregarious behavior, it does show that the animals' bodies were together at a particular point in time. So, it's the trackways that better indicate how they moved and interacted together. As such, scientists have managed to determine that a broad range of dinosaurs did stick together in conspecific groups to forage, nest, or travel. Being careful not to generalize all dinosaurs, it is absolutely true that some dinosaurs of a feather, did indeed flock together.

HOW DO YOU TRAIN YOUR VELOCIRAPTOR LIKE OWEN GRADY?

Owen Grady: "Blue . . . displaying levels of interest, concern, hyperintelligence, cognitive bonding . . . see that? Tilting her head, she's craning forward. Increased eye movement, she's curious. She's showing empathy."
—*Fallen Kingdom* (2018)

Owen Grady. The badass animal behavior expert who has spent more time than most in Jurassic World getting up close and personal with velociraptors. He commands them with a clicker like a pack of dogs, although they're bigger, more vicious, and way more sinister. They can open doors, coordinate attacks, show empathy, and apparently override their brutal instincts to not immediately butcher Grady.

Grady's been working with his raptor pack since they were born, and they've built up a positive relationship with him and each other. But could velociraptors, or any of their close relatives, have been trained to follow his commands? (Please note that in this chapter, velociraptor is mostly in reference to the movie versions, that is, larger dromeosaurs such as deinonychus.)

Animal Instincts

Humans have a history of incorporating animals into our daily activities. The obvious example is dogs, who have lived alongside us for millennia, long before we grew crops or had livestock.

Dogs are descendants of prehistoric wolves and it's thought that humans repeatedly selected them for the most desirable features, leading to increasingly agreeable versions. As such, dogs became the first domesticated animals, followed by sheep, cats, goats, pigs, cattle, chickens, horses, llamas, reindeer, and many others.

Humans domesticated animals for different reasons but mainly to provide food and materials or to work for us. This work included hunting, fighting, defense, transport, and labor in general. To suit our preferences, humans artificially selected the animals based on their observable traits, traits that we now know are encoded within the genes. As such, domestication of animals leads to genetic differences between the wild and domesticated versions and these differences are inherited by the offspring. This process is also called selective breeding and results in various breeds of animals with differences in features such as size, strength, color, pattern, integuments (e.g., hair, skin, feathers, etc.), and also behavior.

Could any of this apply to dinosaurs if they were around now? Well, in an online interview with BuzzFeed.com, paleontologist Stephen Brusatte was asked about raptor training and remarked that "These raptor dinosaurs were very intelligent, had great senses, and were very active, dynamic animals." Adding that, "Their brain size to body size is, in general, kind of like a dog's. So, if you can domesticate dogs, you could probably do it to dinosaurs."

This is reassuring because domesticated animals are relatively harmless to humans when unprovoked. Basically, they're quite chilled and aren't fighting a strong compulsion to rip a chunk out of your throat, as would likely be the case with predatory velociraptors. Domestication shouldn't be confused with taming, though. Taming is specific to an individual animal and involves making it more agreeable to human interaction and control. Taming provides an animal with learned behaviors rather than innate traits, which means a tamed animal won't automatically pass its temperament on to its offspring. This is true of tame lions, tigers, and bears bred in captivity and would also apply to Grady's squad of velociraptors, should they ever breed.

The Raptor Squad

If you recall from a previous chapter, the real prehistoric raptors were likely asocial animals, similar to Komodo dragons, and likely didn't have the social skills for pack hunting. However, the fact that the Jurassic series velociraptors have no problems hunting together suggests that they clearly have a little more going on than your average lizard.

Of the four cloned velociraptors in Jurassic World (Blue, Charlie, Delta, and Echo), Blue is considered the most extraordinary, displaying what Grady describes as "unprecedented levels of compliance." It's clear that Blue exhibited her behavioral differences from an early stage and was potentially born with them. These behaviors make her appear more domesticable, which would make her a prime candidate for a selective breeding program or in the context of Jurassic Park, a reproductive cloning program. It's a wonder why this wasn't an immediate priority, being that she was the last living velociraptor, and especially after she was abducted . . . er . . . "rescued" from the doomed Isla Nublar in *Fallen Kingdom*. Could Blue's unique behavior have a genetic basis, though?

Well, quite possibly. Relatively recent studies have shown that some birds have a cluster of closely related genes that are associated with aggression and parental behavior. One of these genes codes for a hormone called vasoactive intestinal polypeptide (VIP), which increases or decreases aggression or parental care depending on where it is active in the brain. If a similar setup were present in velociraptors, Blue could have had a genetic makeup that predisposed her to being less aggressive or more parental.

The squad was raised together like a family, with Grady playing the part of an adoptive parent or "Mother hen" as Vic Hoskins puts it just before Grady clocks him one in the jaw. Mother hens are people who are regarded as being overprotective or unduly fussy, and considering Grady raised the raptors from hatching, who could blame him if he was? It's obvious he cares for them and has a particular bond that at least Blue clearly reciprocates. So, Grady appears to be like a parental figure to the raptors while also viewing himself as their leader. So, when Gray Mitchell asks him who the alpha of the pack is, Grady responds, "You're looking at him, kid." Does that association hold true though?

The Alpha Issue

The use of the alpha term implies a pecking order or, more broadly, a dominance hierarchy, where someone is at the top while the others are subordinates. Moreover, it was initially believed that alpha dominance was mainly achieved through being the strongest or most vicious in the group, but this view has since changed.

The term alpha was originally based on observations of captive wolves who had often been grouped together artificially from various different packs. It turned out that the observed dominance behaviors in these animals were actually not a true representation of their wild behavior. Instead, wolf packs in the wild tend to be formed from family units, centered around the breeding pair rather than some larger or more able wolves who had vied for dominance. This led to the original alpha explanation being rescinded by its key modern popularizer, David Mech. So, if Grady is the pack's alpha, it means that the pack likely regard him as a dominant member of their family unit, rather than a human that merely dominates them. Although regarding dominance, dog behavior consultant Kayla Fratt explains, "Dominance is far less important in dog social dynamics than previously thought . . . essentially dominance in dogs is related to access to resources, not taking direction from one another. It's much more of a family group than a military hierarchy!" Of course, velociraptors are reptiles, not mammals, but there's no denying that the raptor squad behaves more like a pack of wolves than a bunch of reptiles. In which case, Grady's working relationship with the raptors could have a lot to do with his ability to provide them with resources (such as food and shelter) and so maybe they respond to his leadership because it has consistently led to them accessing these resources.

This might also make sense of the raptor's shifted loyalty when confronted by the Indominus rex. The raptors may have deemed the Indominus as being more able to give them access to resources than Grady, allowing the I-rex to slip into the more dominant position within the group and prompting Grady to declare, "Watch your six. Raptors got a new alpha!" There's also the likelihood that Grady was never the real leader of the squad and that the other raptors were really guided by Blue, a behavior which enabled them to access the resources provided by Grady and his fellow staff. Then when the I-rex appeared, they recognized it as kin and quickly accepted it as a more lucrative alpha than Blue.

So, maybe Grady was never the "alpha" that he thought he was. He just had a unique relationship with Blue, the true leader of the pack.

Follow the Leader

Blue was clearly the one in the bunch that took charge of the other velociraptors. She kept them all in line and provided an example of how they were supposed to behave. So, in a way, she was like a mentor or role model for the raptors. This is something that can be seen in a range of animals, including birds, bearded dragons, and a multitude of mammals, especially dogs. For example, Kayla Fratt points out that, "In some cases, a single well-behaved individual can really help the others behave. This is especially true when working with off-leash dogs on obedience; one or a few well-behaved dogs that come when you call them and/or stay close often will help the other dogs do the same."

These animal role models are more likely to have an effect on members of their own species, and in investigations with humans it's long been known that behaviors are copied more strongly when the role model is similar to the copier. This is a form of observational learning and is hugely dependent on the type of animal and how it relates to the thing it's learning from. As such, there are still many occasions where animals may follow members of another species. This is where imprinting comes in.

There are various type of imprinting, but the one relating to Grady is like filial imprinting. When something is filial, it relates to sons and daughters, meaning that Blue possibly related to Grady as if she were his child. This would assumedly be the same for his relationship with the other raptors, too. So, in essence, he has their trust because they see him as a parent figure, which provides a good basis for training them. Although, as we all know, kids aren't always the best at listening to their parents.

Imprinting occurs within a certain amount of time after hatching, within what's known as a sensitive period. It depends on how developed the animal's senses are but precocial young, which are born ready to go, can imprint sooner. Once imprinted, the young animals can follow their mother figure around and look to that figure to learn how to behave, whether that figure is of the same species or not. There's even been a case of geese imprinting on a pair of wellington boots.

A famous example of imprinting comes from a Canadian called Bill Lishman, who with the help of scientist William Sladen, used an ultralight aircraft to lead imprinted geese on a migration. The 1996 film *Fly Away Home* was based on their exploits. Sometimes Lishman even had them flying alongside him while he rode his motorcycle. So, maybe Grady jumping on his bike to lead a pack of imprinted velociraptors isn't as far a stretch of the imagination as it may at first seem, providing he had sufficiently introduced the velociraptors to the experience of him riding the motorbike beforehand. Although, since they had never been out of containment before, that's unlikely, unless he had already taken his motorbike into their paddock.

Training animals isn't really a matter of imprinting, though. That's just a shortcut to getting them to trust you. And in some cases, it's not so productive for the animal, like when the animal's concept of itself becomes based on the human to the point that it won't even respond to reproductive partners of its own species. Nonetheless, when animal trainers want to get creatures to adopt certain behaviors, there are some particular tried and tested techniques that they use.

Grab the Clicker

We see Grady making extensive use of a clicker to get his raptors to engage, and because many people have a vague idea of a link between clickers and training animals, what he's doing seems totally reasonable. However, this sentiment isn't entirely shared by the people in the know.

Crocodile trainer Colin Stevenson remarks that, "In the movie, the clicker training scene was not realistic, and many in the zoo community laugh at that one. However, it's a movie! I viewed it in the same vein as how there's always a parking space available right where needed in movies." And Kayla Fratt even wrote a whole piece on her Journey Dog Training website about the good and not so good techniques employed by Grady, and how he could have improved them.

Clicker training, and more specifically its underlying principles, has been used successfully on a surprising range of animals, from fleas and fish to birds, mammals, and reptiles. Kayla Fratt has even trained a Komodo dragon, while Colin Stevenson focuses on crocodiles and their kin. "Crocodilians are very responsive to clicker training. In fact, so responsive that clickers are often not used, but simply a command is used. The basic principle is based first on croc behavior and biology, second on basic positive conditioning—rewarding good behavior." So what is clicker training exactly?

Clicker training is a method of affecting the behavior of subjects by almost hijacking their instinctive response to environmental stimuli. The click is used to alert the animal to an upcoming reward, but you have to form this relationship first. This involves giving a reward every time the click is used until the animal associates the click with a reward. This is what's known as Classical or Pavlovian conditioning, named after the nineteenth-century physiologist Ivan Pavlov.

Pavlov realized that his dogs would salivate in the presence of the figure who usually fed them, even if the person didn't have food for them. He developed an experiment to test this observation and ascertained that it was possible to associate a stimulus (e.g. a buzzer) with the dogs' feeding time so that the dogs responded to the buzzer as if they were about to get rewarded (i.e., the dogs salivated upon hearing the buzzer). The dogs had thus been conditioned to associate the sound with the reward.

In classical conditioning, the animal has no control over its response to the stimulus, that is, it's involuntary; however, when the sound is produced by the buzzer (or clicker) it signals to the animal that a reward is coming. Thus, it provides a way to communicate information to an animal based on what the animal values, namely food. So, how do animal trainers use this association to modify behavior?

Animal Training

When the animal exhibits a desired behavior, the click is immediately used to indicate to the animal that the behavior is in line with getting a reward, and a subsequent treat will help to strengthen that notion in the animal. This is an extension of classic conditioning and involves the animal forming an association between their behavior and a particular consequence. This technique is known as operant conditioning (OC).

Operant conditioning was developed by behavioral psychologist B. F. Skinner and involves reinforcing behaviors based on the consequences of reward or punishment. In essence, it makes use of the law of effect, first introduced by Edward Thorndike, who outlined how behaviors with satisfying outcomes are likely to be repeated, while behaviors with dissatisfying outcomes are likely to be reduced. In OC, reinforcement can be provided by giving something satisfying (positive reinforcement) or removing something dissatisfying (negative reinforcement). Alternatively, behavior can be reduced by providing a dissatisfying experience (positive punishment) or removing something they like (negative punishment).

Emily Martin has spent more than a decade training all kinds of animals, including lions, tigers, bears, kangaroos, and birds of prey such as modern raptors! She has firsthand experience of the benefits of operant conditioning and explains its utility in training birds:

> Positive reinforcement is a powerful training tool. [It stems from] finding what the bird sees as reinforcers. For birds that have not been around humans, creating space can start as a reinforcer as you build trust. Food is often a go-to as well, but it's important not to starve your birds. If they decide they do not want to train for their diet, they still get to eat their diet at the end of the day. The zoos I have worked at do not support the use of positive punishment.

Positive punishment is generally harsher, an example of which appears in the very first scene of *Jurassic Park*, where tasers are used in an attempt to stop a velociraptor from mauling an unfortunate gatekeeper. The electric fences around Jurassic Park are also a form of positive punishment. Grady doesn't appear to use any punishment on his beloved raptors, though; their obedience isn't displayed out of fear. It's because he has consistently rewarded them and built up trust as their parent figure.

Putting in the Time

It's vitally important that the process is consistent, as Colin Stevenson reminds us. "Training animals requires short, regular, consistent sessions. If staff cannot maintain this schedule, [or] have a plan for the behaviors they wish to train for, then training is impossible. In fact, things can get worse with inconsistent 'training' sessions that simply confuse and confound the animals."

Using operant conditioning as a basis, simple behaviors can be taught and then combined to build up more complex behaviors. This is known as shaping. However, over time, if the behaviors are not reinforced, it can lead to extinction of the learned behavior, which is something that would have come into play during the three years that Blue was roaming free on Isla Nublar after *Jurassic World*.

In *Jurassic World: Fallen Kingdom*, there's a chance that Blue wouldn't have responded to any specific commands from Grady, instead resorting to the more instinctive relationship that she had with him as her imprinted parental figure. Similar situations have been documented in real life with mammals. For instance, a lion called Christian was able to recognize its previous carers after a year away, and a gorilla called Kwibi affectionately remembered its carer Damian Aspinall after five years in the wild.

The extent to which this relationship holds in reptiles or birds isn't known at the moment, but in the very least, some are known to recognize or distinguish between particular handlers, as described by Emily Martin:

I have seen this a little bit. Birds of prey do not bond like parrots [for example] unless they are imprinted. One of the owls I worked with (not imprinted on any staff at that facility) did act different toward different trainers. She would always work for some and then challenge others. I have seen several raptors act more aggressive toward a certain sex.

Colin Stevenson has also observed similar habits in crocodiles:

There is no question that crocs recognize individual humans. They definitely recognize uniforms of staff and recognize individuals and their voices. This is another thing we rely on: the basic "training" of crocs revolve around using different colored buckets or tools for cleaning as opposed to feeding. The crocs will recognize the pattern quite quickly. They recognize keeper routines as well.

So, where does all of this leave us, in our quest to train our velociraptor?

How to Train Your Velociraptor

Velociraptors are just animals, and humans have tamed and domesticated all sorts for various reasons, so there's no reason to think we couldn't do the same with velociraptors, too, given enough time. Although, a squad working together would probably need some kind of genetic tinkering to make them more amenable to working around each other, considering that some raptors may have actually been asocial animals in real life.

To get your velociraptor to be a team player, it would have needed to be socialized with the rest of the team, whether that's other velociraptors or humans. A way to foster a relationship with humans could be through imprinting, although depending on how complex the genetically tinkered brain was, it could lead to subsequent issues in how it interacts with others of its species, so any other velociraptors intended for use should probably be raised in the same way.

Clicker training would definitely be an initial option for helping your raptor to gain new behaviors, and the clicker would only really be needed early on, with vocal commands or visual cues likely to be just as effective later on. The training should be focused on those behaviors that are necessary for the raptors to function reliably and safely around humans. The training should also be consistent so these behaviors can be built up into more complex interactions.

If you can achieve all of this well, then there's a chance that you too could train a velociraptor like Owen Grady. Although, it helps to be reminded that these would be dangerous animals, not to be taken lightly. So, we'll leave the last words to Colin Stevenson:

> Within the zoo world where we like to think we treat crocs and other animals more realistically and with every respect, we would always maintain safety protocols working with crocs. The training is used to facilitate this, making the work somewhat more predictable and safer, rather than us considering the crocs are "not dangerous." We consider them extremely dangerous and very likely to bite, with training used to minimize the risk and stimulate the crocs at the same time.

HOW DID DINOSAURS DO THE DEED?

Owen Grady: "You might have made them in a test tube, but they don't know that. They're thinking: I gotta eat. I gotta hunt. I gotta . . . you can relate to at least one of those things, right?"
—Owen Grady talking to Claire Dearing, *Jurassic World* (2015)

At the end of *The Lost World*, we see a final clip of a T. rex family and a stegosaurus herd finding their way on the abandoned island of Isla Sorna. The island is full of dinosaurs who are free to do as they please, but there's a problem. The tyrannosaurs are a family of carnivores, all of whom will need to eat. Fortunately for the tyrannosaurs, there's a ready supply of animals on the island. Unfortunately, that supply could be gone within a generation or two, unless their prey start to breed, of course.

In theory, as long as there is a male and female of the species, there's a hope of maintaining the island's population at levels that can sustain a viable food web. In reality, though, whether the dinosaurs actually procreate depends on a number of other factors. For example, is the dinosaur mature enough to actually breed, can it find and attract a worthy suitor, and will they be able to successfully do the deed?

Picturing a situation where Isla Sorna's dinosaur population is giving it a good go at surviving without human interference, what could be said about the dinosaurs' possible breeding antics?

Sex Matters

Dinosaurs had sex; there's no doubt about it. Somehow, the males and females linked up and then nature took its course. However, life has proven itself to be extremely varied when it comes to reproductive unions and looking back from an extremely distant future poses many problems for scientists trying to discern exactly how things might have gone down.

It's highly unlikely that we'd ever find fossilized remains of dinosaurs caught in the middle of the act, but there are other reproduction-related details that can be assessed. These include knowledge about the animals' physical limitations based on biomechanical studies or whether they had any fossilized features that may have been associated with sexual selection, such as elaborate horns, crests, and feathers, collectively known as ornamentation. In some cases, it's even possible to identify whether a dinosaur was a female by observing features that are specific to a particular sex.

For example, ancient bones can be analyzed under microscopes for signs of a special type of bone tissue called medullary bone, which lines the inner bone cavity where the bone marrow is. Medullary bone was first observed in birds but was later found to also be a feature of some of their dinosaur predecessors, particularly theropods such as T. rex. Its presence is unique to females, where the bone is used to store extra calcium, which can later be reabsorbed to provide the minerals needed for building their calcified eggshells. In particular, the production of medullary bone is dependent on estrogen levels, typically related to ovulation, so fossilized dinosaur remains containing medullary bone indicate that the owner was a sexually mature female getting ready to lay eggs.

Another clue to dinosaur reproduction came in 2002 when an oviraptor fossil was discovered in China with a pair of shelled eggs preserved inside its body cavity, meaning the dinosaur remains definitely belonged to a gravid (carrying eggs) female.

The positioning of the eggs suggests they grew from two separate ovaries, something that couldn't have been known by looking at living birds, who only have one functioning ovary. So, there are some fossilized clues to dinosaur reproduction in gravid females at least, although these finds are extremely rare. And it's not just a case of assuming that if birds did it, it must have been the same for their theropod predecessors.

In any case, unless dinosaurs were all reproducing through parthenogenesis, there had to be some preliminary interaction between sexually mature females and males. Generally, this is impossible to determine just from the fossilized remains, so the details must be inferred by drawing analogy with living creatures and assessing the extant phylogenetic bracket (EPB). So, what's the basic outline?

Private Parts

In animals that reproduce sexually, the males and females typically have their own associated reproductive organs, both internally and externally. However, as soft parts don't fossilize well, no one has yet established incontrovertible proof of the exact arrangement of dinosaur dangly bits. Even if we stood face to face with a pair of sexually mature dinosaurs such as the T. rexes in *Lost World*, we'd still have a hard time telling them apart without seeing their related appendages. Although, to be fair, it's hard enough telling the difference between males and females of living species that we interact with on a daily basis. Fortunately, there are many animal specialists who do know about such things, allowing us to learn from the dinosaur's closest living relatives (i.e., crocodilians and birds).

It's known that within the majority of these species, both the females and males have a posterior orifice called a cloaca, which is a single opening used for urination, defecation, and reproduction. It's positioned just under the base of the tail and all reptiles and the vast majority of birds (97 percent) have one so this is probably true of dinosaurs as well. However, to inseminate (introduce sperm into) the female, some of the males also have an intromittent organ (e.g., a penis) tucked away inside their cloacal opening. This is true of crocodilians and the most ancient lineages of birds, such as ratites (e.g., ostriches) and waterfowls, so it's very likely that dinosaurs did too.

For obvious reasons, the exact arrangement of the putative dinosaur penis is still unknown, but if they're anything like modern animals there could be a fair bit of variety among them. For instance, some animals sport a double-headed appendage known as a hemipenis, while others have a prehensile penis that they can fully control. Crocodiles have a member that's in a permanent state of erection, hidden inside their body and springing out when needed before promptly disappearing again. However, the prize for the weirdest appendage goes to the ducks, who have a uniquely spiral-shaped penis that is concealed within them, inside out, but rapidly springs into action when needed. This might seem odd to us, but it has actually evolved to perfectly accommodate the female duck's corkscrew-shaped vagina.

Considering the varied body sizes, shapes, and features across dinosaurs, it's likely that the males' members would have been equally as varied. Evolution literally allows anything that works, no matter how weird it may appear to us. You could imagine a heavily armored ankylosaur having similar problems to a tortoise when it comes to doing the act, and if they evolved a similar solution, an ankylosaur could be well-endowed indeed. Regardless of the male's measurements, dinosaurs obviously had no major problems getting together, considering their lengthy reign on Earth. So, how did they hook up?

Dinosaur Dating

Now, hold your horses! This isn't the part where we go into graphic detail about nasutoceratops nookie. Before anything like that, the dinosaurs need to somehow attract their mate. This involves coming into contact with a member of the opposite sex that is sexually mature (old enough to mate) and also in season (ready to mate).

Regarding the timing of sexual maturity, a 2007 study led by US scientists shed some light on the matter. By analyzing growth lines found within fossilized bones, they found that sexual maturation generally took place before the dinosaurs reached their full adult size, similar to certain living reptiles. This was also apparent in some of the earliest birds; however, extant birds instead grow to full size quickly, then start reproducing later. So, how might sexually mature young dinosaurs have been drawn to each other?

In many species, there are characteristic differences between the forms of the males and females, and this isn't just a reference to their reproductive organs. It's known as sexual dimorphism and involves differences in their external appearance as well as behavior. A classic example is males being a different size than females, although the way the sexes differ isn't universal among all animals. For instance, in some species the males are bigger while in others it's the females that are bigger. It's believed that these characteristics generally evolved as a result of reproductive competition and sexual selection.

Many creatures, including humans, rely on these sexually dimorphic differences as a form of sexual signaling (i.e., to advertise their sexual suitability to members of the opposite sex). It's well-known that many male birds are ornamented with elaborate feathers to attract a female mate, and it's the females' receptiveness to these features that cause them to be passed on to future generations. This is why they are known as sexually selected traits. It's likely that dinosaurs used sexual signaling too, considering the many types of ornamentation they had. Although there has been much debate on whether dinosaur ornamentation was used as a form of sexual display or rather used for other reasons such as species recognition or defense.

The species recognition idea was something supported by Jack Horner back in 2011, although while speaking with *National Geographic* in 2017 he also remarked, "There's no evidence that triceratops used its horns for any kind of aggressive behavior because as they got bigger and bigger, their horns became hollow, and their shields became very thin. They were probably all for display and cause they're displaying very much like birds, more than likely, dinosaurs danced."

Yes, you read it right. Some dinosaurs may have been performing some kind of Mesozoic moonwalk to woo a mate. It had already been established that many dinosaurs had striking colors and patterns on their body and tail feathers. It was also found that some feathered dinosaurs had very flexible and muscular tails that they could have used to shake their tail feathers while attempting to attract a mate. In 2016, US paleontologist Martin Lockley and team analyzed dinosaur tracks and reasoned that the theropod who left them may have been involved in some kind of courtship ritual involving a potential dance, similar to what is seen in modern birds.

Among all this commotion, different sounds may have come into play, too. It's thought that dinosaurs had limited vocal capabilities but still likely made open and closed mouth vocalizations, such as booms and hoots, particularly at lower frequencies where their hearing was best. They may have even made sounds using other parts of their body such as stomping, snapping their jaws together, thumping their tails, or even cracking their tails like whips. In any case, each species would have shown off in its own particular way and its displays would have only appealed to the females of that species.

So, dinosaurs had a lot going on in their dating regime and considering the resurrected park inhabitants wouldn't have had many real world examples to learn from, there's a possibility that they could keep getting the wooing rituals wrong, failing to arouse interest in their potential mate. Additionally, it's known that some animals in captivity just don't breed for various reasons, so there's a chance that this could also be the case for some of the dinosaurs on Isla Sorna.

Nonetheless, we'll assume that the rituals have all been carried out successfully and the dinosaurs have found and attracted their mates. We're finally ready to get down to the nitty-gritty. Now, at this point, things could get a little graphic, so please look away if you're not comfortable with this sort of thing. Actually, who am I kidding?

Doing the Deed

First up, a bit of foreplay. In 2017, paleontologist Thomas Carr and colleagues suggested that tyrannosaur snouts had features that would have made them extremely sensitive. These features were likened to sensors in the skin of crocodilians, called integumentary sensory organs (ISOs). These organs are more sensitive than human fingertips and can detect pressure changes in their environment, kind of like a cat's whiskers but without the hairs. If in this case a tyrannosaur's behavior can be reliably modeled on a crocodilian's, then the scientists propose that alongside utilizing the ISOs to detect temperature and maintain the nest and nestlings, "tyrannosaurids might have rubbed their sensitive faces together as a vital part of pre-copulatory play." Huh, interesting.

Now, aside from the inevitable copulation, nothing's certain about what would happen now. It's mostly shrouded in mystery and some might say it should perhaps stay that way. But, as a usual point of call, to get a vague idea, comparisons have to be made to living creatures. For example, in birds who have cloacas, insemination involves the male lining up its cloacal opening with the female's in what's called a cloacal kiss. The male bird sits atop the female, she moves her tail feathers aside, and the cloacal coitus commences. Unlike birds, though, dinosaurs had huge tails to get in the way, while the males likely had an intromittent organ. Seeing as this is true for crocodiles, too, maybe looking at crocodilian copulation would provide a better example.

Crocodiles usually do it underwater. A male might typically approach from the side and when the female's ready they'll orientate themselves to bring their cloacas into closer alignment. This involves the male getting its leg over the female's back in a "leg over back" posture that is typical of many reptiles. Without a clear view of the insertion point or hands to assist in the docking procedure, you could imagine that intromission (insertion of the penis) would be quite difficult and require a bit of practice to get right. For crocodilians, this is aided by the fact that they do it while submerged with their weight buoyed up by the water. However, crocodilians have a different body structure and manner of walking than dinosaurs did, so there are still limitations as to the applicability of their sex lives to dinosaurs.

So, where does that leave us? Well, in many mammals, the female adopts a lordotic posture in which the front of the body is brought closer to the ground, while the tail end is raised up in the air. The male then mounts the female from behind. This behavior is mirrored in extant large birds such as ratites, where the female responds to the male's signaling to let him know she is receptive, then lowers her body to the ground. The male then assumes the position, sometimes still doing his dating dance along the way (although the actual reproductive act is over in a matter of seconds). It's mostly thought that dinosaurs could have done a similar thing, with the necessity of the females moving their thick, heavy tails into an amenable position.

The sheer size and weight of some dinosaurs poses another problem, although it's quite easy to imagine anything with a weight similar to an elephant getting on fine. The largest land-based carnivorous dinosaurs weighed as much as two large African elephants. Although, as we've seen, dinosaurs were reproducing long before they were full sized, which was probably quite handy given the circumstances. However, some of the larger sauropods weighed more than 10 African elephants, which could have led to problems of potentially bone breaking proportions. Although, it's also been said that since they could support their weight while walking they could probably support the extra weight of a suitor. Even so, the bigger they are, the more the potential for a cock-up, so they would have to conduct themselves in a careful and controlled manner, in any case.

Another well-reported example of careful copulation involves the stegosaurian dinosaur kentrosaurus, which was covered in bony plates and spikes down to the tip of its tail as well as having huge rear-pointing spikes protruding from its shoulders. When scientists used computer simulations to investigate the thorny issue, they concluded that the male would have been castrated if it tried to mount the female from behind, in the usual "leg over back" fashion. As such, they may have opted to do it with one or both parties lying down on their sides, or perhaps the males had an especially adapted organ to allow safe access.

In any case, in a natural environment these creatures would have all had opportunities to see how the deed is done and to practice how to do it themselves. Clearly, any issues we think they had would have been surmountable, considering dinosaurs managed to breed and flourish for so many millions of years. However, the island of Isla Sorna would have presented a unique environment with unnatural circumstances that could have adversely affected their ability to properly function sexually.

Sex on the Island

On Isla Sorna there were various species that would each have had their own specific ways of accomplishing the act, although there's currently no scientific evidence as to how that would have gone down for any of them exactly. What is certain is that it would only have involved the sexually mature members of the island's population, and these dinosaurs would have had to locate a potential partner from that stock. The T. rex tango or stegosaurus salsa would then commence as the male attempts to woo his potential mate while displaying his ornamental features. Whether he gets the dance right or not will determine his chances of passing on his seed. However, without an example of how to do this, he might just fail to excite the female and end up left on the sidelines.

If he does succeed and a potential mate has been wooed, the female would then be receptive to his advances and the male would line up to establish the union. Although again, without any prior practice this could prove slightly problematic to get right at first. Given enough time, they'd probably stumble across solutions eventually but with the limited island population it's also very possible that they just wouldn't get the opportunity. Worse still, they might just not feel the urge to breed, as is often the case with animals bred in captivity.

Mesozoic dinosaurs were free to do the deed in the wild and became very effective at it despite their many perceived obstacles. However, it's a mystery whether the ins and outs of the process would be easily accomplished by a marginalized island population of virgins, who were mostly female. If successful, long may they prosper. If unsuccessful, then they'd probably just live out their lives with limited subsequent breeding cycles before the island's population collapsed. Some or all of the dinosaurs would eventually die out and the island would return to supporting its native population of species, alongside any particularly successful new additions.

FALLEN KINGDOM: HOW WOULD TRUMP'S AMERICA DEAL WITH THE ESCAPED DINOSAURS?

Ian Malcolm (voiceover): "We're causing our own extinction. Too many red lines have been crossed. And our home has, in fundamental ways, been polluted by avarice and political megalomania. Genetic power has now been unleashed. And of course, that's gonna be catastrophic. This change was inevitable, from the moment we brought the first dinosaur back from extinction. We convince ourselves that sudden change is something that happens outside the normal order of things, like a car crash. Or that it's beyond our control, like a fatal illness. We don't conceive of sudden, radical, irrational change as woven into the very fabric of existence. Yet, I can assure you, it most assuredly is. And it's happening now. Humans and dinosaurs are now gonna be forced to coexist. These creatures were here before us. And if we're not careful, they're gonna be here after. We're gonna have to adjust to new threats that we can't imagine. We've entered a new era. Welcome, to Jurassic World."

—D. Connolly and C. Trevorrow,
Jurassic World: Fallen Kingdom (2018)

"Today, around the world, demagogues appeal to our worst instincts. Conspiracy theories once confined to the fringe are going mainstream. It's as if the age of reason, the era of evidential argument, is ending, and our knowledge is increasingly delegitimized and scientific consensus is dismissed. Democracy, which depends on shared truths, is in retreat, and autocracy, which depends on shared lies, is on the march. Fake news outperforms real news. On the Internet, everything can appear equally legitimate. The rantings of a lunatic can seem as credible as a Nobel Prize winner."

—Sacha Baron-Cohen, ADL International Leadership Award speech (2019)

All of Our Dinosaurs Are Missing

As *Jurassic World: Fallen Kingdom* took over $1.3 billion at the box office, another sequel was all but guaranteed. Steven Spielberg confirmed that *Jurassic World 3* (or *Jurassic Park 6*, depending on how one views the franchise) would soon be upon us. So how will the three-quel follow on from *Fallen Kingdom*'s cliffhanger, which saw dinosaurs released into the world?

Readers will recall the end of *Fallen Kingdom*. Eli Mills's plan to auction off the captured dinosaurs on the black market ends in chaos, as Maisie Lockwood ultimately frees the caged dinosaurs into Northern California. At least two dozen different species of dinosaur are liberated. The escapees include a number of familiar species, such as brachiosaurus, apatosaurus, and triceratops, and perhaps the more notably dangerous allosauruses, T. rex, and velociraptor, the latter being Blue, the raptor trained by Owen Grady in the Jurassic World theme park. One of the film's closing scenes is a new US Senate hearing in which Dr. Ian Malcolm declares the dawning of a Neo-Jurassic Age. Now, humans and dinosaurs must coexist. The movie ends as we witness the discharged dinosaurs roaming wilderness and urban areas alike. What's not mentioned in the movie, of course, is that the dinosaurs have been set free at a very particular time: this is Trump's America. Can real life tell us anything about how a fictional President might greet the prospect of an invasion of immigrant dinosaurs?

President Denies Dinosaurs Escaped

So, the island has blown up. The beasts are out of the box. What happens next? Surely the story won't turn into a dinosaur-versus-human disaster movie . . . will it? There are responsible people in power. But in the early days of what would turn out to be a massacre, those in political power seek to minimize the sense of threat, as they do with pandemics. Professor Tony Faucet, a Nobel Prize–winning dinosaur expert, warns about the genetic power that's been unleashed. But the President even denies that there are any escaped dinosaurs. He tweets, "We have it very well under control. We have very little problem in this country at this moment. We're working very closely with China, and it's going to have a very good ending for us, that I can assure you."

President Predicts Dinosaurs Won't Attack

The President seems amazed when he privately tells a journalist that many of the dinosaurs are apparently "deadly," but nonetheless continues to inform his devoted public that all is "under control," and even that they could go away in the spring. The President predicts that the beasts are not naturally violent, as an increasing number of reported dinosaur attacks flood in from across the country. When confronted with the facts, he tweets, "I predict the dinosaurs will go away in April. We're in great shape, though. We have only a dozen cases, and many of the victims who are not dead are doing very well once their limbs have been sewn back on. We're in good shape now."

President "Knew Dinosaurs Escaped Before Anyone Else"

When it becomes clear that there is a huge surge in the number of dinosaur attacks, the President tweets, "Of course, no one knows more about dinosaurs than me. Throughout my life, my two greatest assets have been mental stability and being, like, really smart. I went to the best colleges for college. My very good brain told me these dinosaurs had escaped before anyone knew." Later, in an attempt to underplay the scale of the massacre, the President says, "I want you to know something that shocked me. I spoke with Faucet, that Nobel Prize–winning scientist guy, right? I mean, what does *he* know that I don't? But he did tell me this—the flu, in our country, kills between 25,000 and 69,000 people a year. Over the last 10 years, we've lost 360,000. That was shocking to me. That's far worse than dinosaur deaths."

President Slams "Chinese Dinosaurs"

Soon, the dinosaur massacre is nationwide. Those in power take steps to slow the carnage, with public events being banned. Nonetheless, the President talks about soon reopening and having "packed churches all over our country." "My administration has taken the most aggressive action in modern history to protect Americans from the dinosaurs," says the President, "and most of the 100,000 deaths are largely old people who can't run away. But I've always known this is a—this is a real—this is a massacre. I've felt it was a massacre long before it was called a massacre. I've always taken these Chinese dinosaurs very seriously. Yes, *Chinese* dinosaurs. That's not racist at all. They come from China, that's why. The real big meat-eater? It comes from China. I want to be accurate."

President Claims "No One Could Have Predicted the Dinosaurs"

Reports come in of churchgoers being eaten alive during lock-down-busting services. News of the attempted mass shootings of dinosaurs is everywhere. With gun control more of an issue than ever, the President turns to the question of novel forms of protection from dinosaurs. "Don't listen to the scare-mongering scientists, basing their opinions on mere facts. Suppose we hit the dinosaur body with a tremendous—whether it's ultraviolet or just very powerful light? Get that light inside the body somehow, either through their incredibly thick skin or in some other way. Or maybe strong disinfectant will knock the dinosaur out in a minute. One minute. By blow dart or by injection inside or almost a cleaning? If you can get close enough. I hear that Raptor can move pretty fast, but not as fast as me. Some scientist should check the disinfectant. We have to think on our feet as no one could have predicted these dinosaurs."

President Says Dinosaur Woman Man Camera TV

As a second wave of dinosaur deaths seems imminent, the conspiracy group R-Nameless organize a series of rallies across the country claiming the entire dinosaur invasion is a hoax. A spokesperson for R-Nameless opines on camera, "The whole thing is a bunch of hooey. First the scientists say the dinosaurs are not dangerous, then they say they are! Why don't scientists make up their minds?" Before the interviewer can point out the difference between a brachiosaurus and a T. rex, the R-Nameless spokesperson is bitten down to the waist by Blue. As hundreds of conspiracists at the edge of the throng are picked off by the T. rex and allosauruses, the rest of the crowd chant, "Down with the liberal dinosaur hoax!"

Dinosaurs with Guns

The second wave hits a peak when the allosauruses finally find a use for their relatively puny arms—guns! They arm themselves with AR-15s all too easily acquired from Walmart, as the T. rex tools up with a .500 Nitro Express (ironically known to humans as a T. rex killer). Some state police are powerless to stop the armed and marauding creatures, as open carry is law. The death rate becomes exponential.

Media Spreading "Fake News" about Dinosaurs

It becomes clear that the dinosaurs have taken more American lives in a few months than were lost fighting every single war since 1945. A major television documentary attempts to expose the way in which the President has sought to sabotage a more scientific program to keep people safe: purposefully misrepresenting the threat early on; belittling PPE; encouraging people to protest against lockdown rules; anything to undermine, suppress, and censor scientists working to reduce the harm from dinosaur attacks. When confronted with these facts in an interview at a besieged White House, the President tells an interviewer, "The fake news is talking about CASES, CASES, CASES. This includes many low-risk people, people who've already had their limbs bitten off. Media is doing everything in its power to make me look bad. The cases are up because reporting is way up by fake news agencies." As the T. rex thunders into the White House and eats the President whole, the interviewer concludes, "and there we have it, ladies and gentlemen, natural selection in action."

WAS MOSASAURUS THE GREATEST SEA BEAST?

"The Mosasaurus was thought to have hunted near the surface of the water where it preyed on anything it could sink its teeth into. Including turtles, large fish, even smaller mosasaurs."

—Announcer, *Jurassic World* (2015)

Who could forget that jaw-dropping moment when the Mosasaurus breaches the water's surface to grab a suspended shark? The sheer size of the creature astounds as its gargantuan jaws snap shut around the great white and the beast crashes back into the water, splashing the crowd. It's Sea World on steroids, but this aquatic leviathan is up to 83 million years displaced in time.

Could any other sea beast trump these monstrous Mesozoic marvels? Well, to find out, let's check the stats, stack the facts, and measure Mosasaurus up against some of the other biggest and most ferocious creatures to have ever swam the seas.

Fact File 1: Mosasaurus

The Mosasaurus we see on the screen is quite simply massive, although just how big it is turns out to be hard to pin down. Depending on where you get the information, it could measure anywhere between 55 feet and 85 feet long, while weighing in at more than 61,000 pounds, the equivalent of three tyrannosaurus rexes.

It's kept in a lagoon that's purported to hold 3 million gallons of water. However, that number seems woefully undersized. For comparison, the largest tank (the show pool) at Sea World holds 2.2 million gallons of water in a region that's about 36 feet deep and

165 feet across. The combined amount of water in all of the orca tanks is 6 million gallons, although the up-to-22-feet long orcas can't move freely between these tanks. There are some groups that regard these habitats as too small and artificial a home for orcas to occupy permanently, so at 2.5 to 4 times bigger than those orcas, the massive movie Mosasaurus would have needed a substantially bigger space.

Real Mosasauruses weren't that big, though. Of the few species of Mosasaurus currently recognized by scientists, the biggest is Mosasaurus hoffmannii. Its size had to be extrapolated from the dimensions of its fossilized skull but estimates range from about 36 feet to 55 feet long. This means that the biggest estimates of a real Mosasaurus tallies with the smallest measurements given in movie canon (i.e., 55 feet). So, it's possible that some of the in-movie stats about the size of Mosasaurus may have been referring to the real discoveries of Mosasaurus remains, rather than the freak Mosasaurus held in the Jurassic World tank.

Mosasaurus was the first discovered example of a group of lizards called the mosasaurs, which marauded the seas of the Late Cretaceous Period from about 100 million years ago until their demise in the cretaceous extinction event. Being reptiles, they breathed air so they had to return to the surface to catch breaths and considering that mosasaurs appeared to be surface dwellers rather than deep sea divers, they may not have felt so out of place in the relatively shallow habitat provided them in Jurassic World.

Overall, the mosasaurs were pretty successful, with fossils being found in many parts of the world. They lived alongside other predators like crocodiles, sharks, and plesiosaurs, some of which would have been on their menu alongside fish, sea turtles, and other mosasaurs. So, mosasaurs were likely near the top of their food chain.

The mosasaurs were part of an order of scaled reptiles known as squamates, making them close relatives of snakes and monitor lizards (such as Komodo dragons), which explains some of their physical features. They had huge, powerful, double-hinged jaws with an extra set of teeth in the roof of their mouths (called pterygoid teeth). A 2005 study led by Dutch researchers has even suggested that they may have had forked tongues. Various fossilized skin remains have revealed that some of them did indeed have a scaly and snake-like covering while the color of the mosasaur has also been found through analysis of microscopic, fossilized structures called melanosomes. The mosasaurs had strong tails, which they used to power them through the water, and a recent study suggests that they may have also used their forelimbs in a breaststroke fashion to help propel them along.

Now, if you wanted to catch a real-life glimpse of a gigantic mosasaur skeleton, you could check out another group of large mosasaurs known as tylosaurs. The Canadian Fossil Discovery Centre (CFDC) has a 42-foot tylosaur, nicknamed Bruce, which is the largest publicly displayed mosasaur in the world. In 2009, staff at the CFDC also found the jaws of a xiphactinus clamped onto the flipper of a mosasaur. At lengths of up to 20 feet, the carnivorous xiphactinus was one of the largest bony fish of the time. Nonetheless, mosasaurs are regarded as being a ruling class of marine reptiles during the Late Cretaceous.

But just because mosasaurs ruled in one part of the Mesozoic, it doesn't make them the greatest sea beast. If we look earlier in the Mesozoic, there were many other predatory reptiles that would have challenged its status.

Fact File 2: Marine Reptiles

The Mesozoic Era has been labeled the "Age of Reptiles," a time when reptiles diversified to dominate the land, air, and sea. The earliest marine reptiles included well-known sea creatures like ichthyosaurs from the early Triassic, 250 million years ago, and plesiosaurs from the late Triassic, 200 million years ago. The

Triassic also saw other predatory marine reptiles that died out soon afterward such as the crocodile-like phytosaurs or the thalattosaurs, whose name means "ocean lizards."

The ichthyosaurs lasted until the mid-Cretaceouss and had a body form that was similar to dolphins, which appeared much later and are genetically unrelated. The icthyosaurs included many top predators that fed on cephalopods, fish, and other ichthyosaurs. One 240-million-year-old specimen (described as a Triassic mega-predator) was even found with the torso of a 13-foot thalattosaur in its stomach. Although, some of the largest ichthyosaurs were the shastasaurids of the Late Triassic, ranging between 20 feet and 66 feet long. In 2018, UK scientists reported on partial remains of an ichthyosaur found in Lilstock, UK, that indicate it may have been between 66 feet and 82 feet long, making it comparable in size to blue whales and bigger than the largest carnivorous dinosaur, Spinosaurus.

In the past, it was claimed that some of the larger dinosaurs were probably aquatic, thinking that they may not have been able to support their massive body weights otherwise. This view is now largely discredited, as multiple lines of evidence (such as body shape, bone density, tail flexibility, and nostril and eye location) have shown that dinosaurs most definitely were land based. Although, the fish-eating spinosaurs were always a possible exception, with their long snouts and conical fish-grabbing teeth.

Spinosaurus is estimated at around 50 feet long and lived in the Early Cretaceous of what is now North Africa. A Spinosaurus appears in *Jurassic Park III*, patrolling the land and hunting humans, then later swimming through the water to stalk the humans on their boat. Semi-aquatic behavior was long suspected of this dinosaur, and in 2020, researchers reported details from fossilized tail remains that provided more evidence of this ability. It appeared that Spinosaurus had a paddle-shaped tail, which biomechanical models indicate would have helped it to be a faster and more efficient swimmer. The lead researcher, Nizar Ibrahim, described it as, "basically a dinosaur trying to build a fish tail." One species of spinosaur called baryonyx was even found with partially digested fish scales in its stomach, although there were also iguanadon bones, so its diet was likely quite varied.

Remains of the last meal in the stomach is the most direct way of assessing what the Mesozoic marine reptiles dined on. As yet, only five genera (of Mesozoic marine reptiles) have been found with the remains of other tetrapods (four-limbed creatures) inside them. These genera were all from two groups, the mosasaurs and the ichthyosaurs. However, another bunch of large marine predators worth a mention are the plesiosaurs and pliosaurs, which are sometimes confusingly bunched together under the banner plesiosaurs, or more specifically, Plesiosauria, which is the term we'll use if referring to both groups here.

The plesiosaurs survived until the end of the Cretaceous and can be recognized by the general body plan of a long neck and small head (i.e., the archetypal Loch Ness monster), with limbs comprised of four paddle-like flippers, which they moved in a flapping motion as if flying through the water. Along with mosasaurs, plesiosaurs helped to fill some of the niches left open after the ichthyosaurs and pliosaurs went extinct around 90 million years ago. The pliosaurs had the same flipper arrangement as plesiosaurs, although the body plan generally featured a short powerful neck and large head. Examples of these apex predators include the 20-foot mostly Late Jurassic liopleurodon, the 33-foot pliosaurus funkei (which had an estimated bite force of up to 49,000 newtons and was temporarily dubbed predator X before being identified and named), or the similarly sized kronosaurus of the Cretaceous Period.

Many of these beastly marine reptiles were comparable in size to real Mosasauruses, with larger ichthyosaurs and many ferocious contenders among the Plesiosauria. However, beyond the Mesozoic, some of the greatest sea beasts were not reptiles at all.

Fact File 3: Great Sea Beasts

Many of the biggest sea creatures to ever have existed were actually fish and mammals. If we go back more than 350 million years to the Late Devonian Period, there's an almost 30-foot-long armored fish called dunkleosteus. This marine predator had a more than 6,000 newton bite force, capable of breaking through the hard shells of underwater prey, including other dunkleosteuses.

The Devonian was known as "the age of fishes"; however, the biggest fish that ever lived was the approximately 50-foot-long leedsichthys, which swam the Jurassic seas 160 million years ago. These were filter feeders that mainly fed on plankton so, although they were enormous, they weren't predators and were likely preyed upon by liopleurodon. In comparison, the biggest living fish is the whale shark, weighing over 44,000 pounds and measuring more than 40 feet in length, with some potentially achieving lengths of more than 60 feet. Like leedsichthys, whale sharks are filter feeders and not predators, which is also the case for the largest sea creature that has ever existed, the blue whale.

The largest known blue whales spanned 110 feet and weighed 420 thousand pounds. In fact, the largest living creatures are all whales. If you didn't know already, whales are actually mammals, and just like marine reptiles, they originally evolved on land before returning to the sea, becoming increasingly adapted to an aquatic lifestyle. As such, they have to return to the surface to take in lungfuls of air, as their respiratory system is still configured for air breathing, like all of their mammalian and terrestrial relatives.

In 2018, scientists at Stanford University looked into the size distribution of aquatic mammals, reporting that "the aquatic realm imposes stronger constraints on body size than does the terrestrial realm, driving and confining aquatic mammals to larger sizes." It turns out that they got so big as an advantage toward retaining body heat, which in turn increased their metabolism, requiring more food to be consumed. Additionally, with their higher levels of body fat to keep them warm, whales could migrate into colder seas that couldn't be survived by the less blubbery marine predators that stalked them.

There is one well-known extant sea creature that does not have any predators, though. The Orcinus orca, also known by the apt title "killer whale." Orcas can measure more than 30 feet and weigh up to 22 thousand pounds. Despite their name, they aren't actually whales, but instead are part of the dolphin family. Known to hunt in groups, they have various prey including cephalopods, birds, other sea mammals, and various fish, including that other well-known apex predator of the sea, great white sharks. This positions orcas as the most fearsome mammalian carnivores of the sea, as well as making them the biggest extant apex predator on the planet.

Not to take anything away from the awesomely fierce great white shark, though, which is in fact a large species of cartilaginous fish. The great white shark is the largest and most ferocious living predatory fish, with the biggest measuring 20 feet and weighing over 5,500 pounds. They have a potential bite force of more than 18 thousand newtons, which is in the region of a saltwater crocodile's bite, which in 2012 was the strongest bite ever directly measured. (Although, this is still much less than the bite force estimated for large pliosaurs.) Great whites weren't the biggest predatory sharks that ever lived, though. That title goes to the spectacular Megalodon, whose name basically means "large tooth."

Megalodon lived from roughly 23 million to 4 million years ago, during the Neogene Period. It's estimated that they could have been up to 59 feet long, which is three times longer than a great white. But, while a great white's teeth are about 1.5 inches in length, megalodon teeth were about 7 inches long and four times wider (hence the name "large tooth"), although it should be noted that the size of a megalodon is generally extrapolated from the size of its teeth. Nonetheless, it's been estimated that megalodon had a bite force of between 108,000 and 180,000 newtons, which is between six and ten times stronger than a great white shark's bite. Megalodon is considered the largest carnivorous fish that ever lived. A 2020 study led by researchers at the University of Bristol and Swansea University found that a 52.5-foot otodus megalodon would have a 5.3-foot dorsal fin (on its back), with a 15.3-foot head, and a tail measuring 12.6 feet high.

Despite their impressive stats, the megalodons succumbed to extinction and it's thought that climatic changes may have played a major part. As the surface waters got increasingly cooler, their major food sources such as whales could survive due to their extra layers of fat, but without equivalent levels of insulation, the megalodons were constrained to the warmer seas, with smaller prey that couldn't support their massive size. A study published in 2019 also noted that they seem to have gone extinct around the same time as the appearance of great white sharks, indicating that the great whites may have outcompeted the megalodons. In particular, the juvenile megalodons would have been of comparable size to adult great whites (i.e., they shared the same niche and thus competed for the same food).

So, Was Mosasaurus the Greatest Sea Beast?

Mosasaurus was just one of the many great sea beasts that existed on this planet. They most certainly weren't the biggest sea beasts, though, being outdone by other gigantic reptiles, fish, and mammals that roamed the seas at different points in Earth's history.

Mosasaurs weren't the toughest either, being overshadowed by predators such as megalodon. However, if a Mosasaurus was resurrected to have the proportions of the one seen in *Jurassic World*, there would probably be no predatory contest. Similar to how beyond a certain size the great land dinosaurs were pretty much untouchable. Although, such a mosasaur would very likely struggle to get in enough food to sustain it, considering that the most successful gigantic sea creatures tended to be filter feeders.

So, even though the Mosasaurus escaped confinement in *Jurassic World: Fallen Kingdom*, it's very possible that this last specimen would have soon succumbed to hunger in a similar way to the megalodon. While individually it's a great and formidable beast, its realistic survival away from captivity would probably require it to be much smaller than the mammoth movie monster and maybe more like the specimens seen in the fossil record. I'll leave the last words to marine biologist Tom "Blowfish" Hird (who just informed me of another awesome sea beast called Livyatan melvillei!). "It seems then that the only place where mosasaurus would really hold the title of greatest sea beast is in its own wildest dreams, rather than our current modern seas."

SHOULD WE MAKE A REAL-LIFE JURASSIC WORLD?

"Scientists are actually preoccupied with accomplishment. So, they are focused on whether they *can* do something. They never stop to ask if they *should* do something. They conveniently define such considerations as pointless. If they don't do it, someone else will. Discovery, they believe, is inevitable. So, they just try to do it first. That's the game in science. Even pure scientific discovery is an aggressive, penetrative act. It takes big equipment, and it literally changes the world afterward. Particle accelerators scar the land and leave radioactive by-products. Astronauts leave trash on the moon. There is always some proof that scientists were there, making their discoveries. Discovery is always a rape of the natural world. Always."

—Michael Crichton, *Jurassic Park* (1990)

Lost Worlds

The Jurassic World franchise is a new and more scientific form of the old and romantic "lost world" tales. Until science fiction came along, lost worlds were unheard of. A typical lost world tale would first find our adventurer somewhere in the civilized world, usually London (these tales were often written at the height of British imperialism). Armed with a tall story or an ancient scroll, our hero sets off to unknown lands and lost civilizations to find secret powers of great antiquity. The quintessential lost world was Plato's Atlantis. But the classic lost world stories of the Victorian age, such as H. Rider Haggard's *King Solomon's Mines* and *She: A History of Adventure* (the famous "She Who Must Be Obeyed") are with us still. Such stories survive in video game adventures like the Tomb Raider series with Lara Croft (she, too, must be obeyed).

First came the exploration and exploitation of the natural world. After all, before you can go find a lost world, it's got to go missing in the first place. The medieval voyages of discovery had really opened up our planet to piracy and plunder. One of the main architects of the new philosophy of medieval science was an English statesman called Francis Bacon. Bacon was the key prophet and publicist of the new age, a kind of Renaissance spin doctor. He believed that organized science would forge material progress and seized on the idea that understanding nature was the main means of taming the planet for profit. He wasn't wrong.

In an echo of Michael Crichton's remarks about "ungovernable science," but many centuries before, Bacon foresaw a utopia, and empire, of science. In his diaries, he claimed to be seeking to "enlarge the bounds of human empire to make all things possible." This desire for power, however, violently crushed alternatives.

Bacon believed that nothing should get in the way of progress and, to this end, he developed a flawed medievalist ideology of power with his idea of "monstrosity." There were peoples, Bacon said, who had degenerated from the laws of nature, and become monstrous. Among such multitudes, who Bacon believed deserved destruction were (according to Bacon's diaries), "West Indians, Canaanites, pirates, land rovers, assassins, Amazons, and Anabaptists." Armed with this kind of repugnant philosophy, the expropriating British Empire expanded. Native colonial peoples dispossessed, shot, poisoned, and diseased.

Worlds were, essentially, lost. Before the 1770s, large parts of the world remained unknown to Europeans. As science and technology grew, and as modern "civilization" crept around the globe, fantastic travelers' tales became very popular. One of the very first examples was Ludvig Holberg's *Nicolaii Klimii iter Subterraneum* (1745). Translated as *A Journey to the World Underground*, or simply *Niels Klim's Underground Travels*, the tale tells of a young Norwegian who stumbles down into the Earth to discover an inner planet populated by intelligent non-human life forms.

Perhaps a more familiar take on such a lost world story is Jules Verne's *Journey to the Center of the Earth*, written in 1864. Verne's creative journey had begun in 1863 with the first of 63 *Voyages Extraordinaires: Voyages in Known and Unknown Worlds*. An early advertisement claimed that Verne's goal was "to outline all the geographical, geological, physical, and astronomical knowledge amassed by modern science and to recount, in an entertaining and picturesque format that is his own, the history of the universe." Some mission.

However, whereas *Journey to the Center of the Earth* is classic Jules Verne science fiction, he was quite unapologetic about the penetrative thrust of science into nature. Verne's book is a voyage through a subterranean world, and a conquest of space. The novel's main idea is this: nature is a cypher to be cracked. It's a journey into the depths of evolutionary time. Once into the Earth's subterranean caverns, grottos, and waters, Verne's explorers find the interior alive with prehistoric plant and animal life. A herd of mastodons, giant insects, and a deadly battle between an ichthyosaurus and a plesiosaurus follow. Verne's book promotes a giddy confidence in progress, and a predictable cosmos in which the unknown is easily assimilated into our taxonomies.

So, science had led to a kind of creeping separation from nature. The sheer pace of dizzying progress accelerated in Victorian times when the emergent sciences of biology and geology made the modern feeling of alienation even worse. Science fiction began to try to repair the separation from nature, to reload the emptiness, to somehow jack into the void. Verne's fiction finds lost worlds and dinosaurs by exploring geographical space. Michael Crichton's *Jurassic Park* did the same by jacking into our genetic past.

The Gene Genie

It's hardly surprising that we get excited by the prospect that, someday, the dinosaurs will live again. Look at what genetics has done for human history. Gene markers in human blood are a type of time machine. Using these markers, scientists can look back into Earth's history, as inside a drop of blood is the best history book ever written, with everyone on Earth carrying a unique chapter in their veins. We've learned how to read the time machine in our genes by taking blood samples from people all around the planet. The conclusions are stunning. All humans alive today are related by blood, in one big family tree. It really wasn't that long ago that there were only about two thousand humans, living in a single continent of Africa. Our blood tells the tale of a small group of ancient humans who left Africa on a long journey. Those of us alive today are their children, and we are still working on the story of how our ancestors came to populate the Earth in surely the most incredible journey in human history.

What if we could use genetics to recover lost worlds? In *Jurassic Park*, Michael Crichton was ahead of the curve. To bring dinosaurs back from the dead would require an intact genome, notoriously hard to find. You'd need to find a genome template from an ancient sample with sufficient data to make more than just a few genes. Tricky. Also, the quality and quantity of prehistoric DNA would depend on the preservation and degradation of the sample's DNA. True, DNA is surely stable enough for nature to trust it through evolutionary time. Nonetheless, DNA *does* degrade over time. And so, the challenge would be finding enough intact pieces in prehistoric DNA, especially in amber-trapped insects found in South American rain forest habitats. There might not even be enough intact DNA to clone a single deadly raptor claw. And yet the point was that Crichton fired our dreams. He cloned dinosaurs. He let the imagination of science soar and raised expectations of new and fantastical future discovery.

Science has recently caught up with science fiction. Recent advances in genetic tech have boosted the hope of being able to clone prehistoric creatures in the future. Scientists have successfully sequenced a 10,000-year-old woolly mammoth, a 38,000-year-old Neanderthal, and the 80,000-year-old genome of a young female of an early species of homo sapiens called the Denisovans—a close relation to the Neanderthals. The project team even reported the Denisovan girl had brown skin, eyes, and hair. Finally, pushing back the barriers even further into our prehistoric past, the entire genetic sequence of a 700,000-year-old extinct species of horse was recently published in the journal *Nature*.

A final important aspect of Crichton's science fiction remains, and that's the question of chaos theory. It's encapsulated in the words of Crichton's lead character, Dr. Ian Malcolm, the skeptical chaotician in the fiction and film:

> Science cannot help us decide what to do with that world, or how to live. Science can make a nuclear reactor, but it cannot tell us not to build it. Science can make pesticide but cannot tell us not to use it. And our world starts to seem polluted in fundamental ways—air and water and land—because of ungovernable science.

One solution may be in taking such decisions out of the hands of private capital, and into the public arena. To ensure that future science projects don't simply count down the seconds to disaster, we the people need to make the decisions.

EXTINCTION: WILL HUMANS GO THE WAY OF THE DINOSAURS?

"Of the four billion life forms which have existed on this planet, three billion, nine hundred and sixty million are now extinct. We don't know why. Some by wanton extinction, some through natural catastrophe, some destroyed by meteorites and asteroids. In the light of these mass extinctions, it really does seem unreasonable to suppose that homo sapiens should be exempt. Our species will have been one of the shortest-lived of all, a mere blink, you may say, in the eye of time."
— P. D. James, *The Children of Men* (1992)

"The planet has been through a lot worse than us. Been through earthquakes, volcanoes, plate tectonics, continental drift, solar flares, sunspots, magnetic storms, the magnetic reversal of the poles . . . hundreds of thousands of years of bombardment by comets and asteroids and meteors, worldwide floods, tidal waves, worldwide fires, erosion, cosmic rays, recurring ice ages . . . and we think some plastic bags and some aluminum cans are going to make a difference? The planet isn't going anywhere. *We* are! We're going away. Pack your shit, folks. We're going away. And we won't leave much of a trace, either. Maybe a little Styrofoam. The planet'll be here and we'll be long gone. Just another failed mutation. Just another closed-end biological mistake. An evolutionary cul-de-sac. The planet'll shake us off like a bad case of fleas."
— George Carlin, *The Planet is Fine* (1992)

Extinction Now!

In *Jurassic World: Fallen Kingdom* there is an agency called "Extinction Now!" The organization acts as a mouthpiece to protest against the imminent extinction of the remaining dinosaurs on Isla Nublar. Extinction Now! appears to have a similar point of view to Dr. Ian Malcolm's as they quote him on their website. (Incidentally, as part of the PR for the launch of the movie, an actual website was created for Extinction Now! that was launched shortly before the movie's release.)

The dinosaur extinction on Isla Nublar, of course, would be due to the imminent eruption of the active volcano Mount Sibo in the north of the island. But what of human extinction? In the movie, Ian Malcolm tells a US Senate hearing, "How many times do you have to see the evidence? We're causing our own extinction. Too many red lines have been crossed. And our home has, in fundamental ways, been polluted by avarice and political megalomania." Does the good Dr. Malcolm have a point: Will the human species go extinct?

Extinction When?

According to conventional science, the short answer is, yes, very likely. It is a basic principle of paleontology that all species go extinct. Indeed, the fossil record shows that over 99.9 percent of all species that have ever lived are now extinct. Naturally, some species left successors. But brontosaurus and trilobites didn't. Neither did fellow human species Neanderthals, Denisovans, and homo erectus. Only homo sapiens remain extant. Are we human successors also inevitably heading for extinction? Or can we cheat our imminent mass death?

Online clickbait often wants us to believe that this extinction is almost upon us. Perhaps it's the threat of an Earth-crashing comet, like the one that did in the dinosaurs. That's often a social media favorite. Then there's the ongoing threat of global warming and the climate emergency. And they have a point, of course. We humans are vulnerable, warm-blooded creatures that don't deal with ecological disruptions very well. Smaller, cold-blooded creatures, such as turtles and snakes, can last for months without food, so are better equipped for survival. But bigger beasts with faster metabolisms such as humans and Tyrannosaurs need lots of food. Nonstop. And that leaves us prone to even brief food-chain disruptions, such as those caused by pandemics, or catastrophes such as volcanoes, mini ice ages, or the nuclear winter that happens after a cometary collision.

There are other unexpected factors that put us in danger. We live longer. We have longer generation times. And we have few offspring. Such a relatively slow reproduction rate makes it harder for us to recover when the population bombs, and also makes it harder to deal with breakneck environmental changes. These are the kinds of factors that did in doomed megafauna such as mammoths and ground sloths. These large mammals reproduced far too slowly to adapt to being overhunted by humans.

Go Forth and Multiply!

So, humans are endangered. Naturally, it's tempting to be seduced by the idea that humans are resistant to extinction. Perhaps uniquely so. After all, surely, we're a profoundly unique species? We are widespread. We are abundant. And we are peerlessly adaptable. All traits that suggest we'll survive a future crisis.

For one thing, we're global. Back in 1859 when Charles Darwin wrote *Origin of Species*, among many other matters he identified the mechanisms of natural selection. Mechanism one, "multiplicity," is good. The idea that a species should make more offspring than the environment can necessarily contain. Mechanism two, "geographical spread," in which the farther afield a species can spread, the less tied it is to a single setting. And just look at us humans. We're not just everywhere, we're superabundant. A world population just shy of 8 billion souls. We're among the most common creatures on the planet. Human biomass is greater than that of all wild mammals. Even if a pandemic or nuclear war killed off 99 percent of humans, millions would still survive to wipe the slate clean and reboot.

And, hell, we're geographically widespread. As organisms, we should fare better during catastrophes such as a comet impact or mass extinction events. If one habitat is annihilated, we can survive in another. Polar bears and pandas would be done for. It's the small range of habitat that would kill them. But brown bears and red foxes would fare better. With much larger ranges, they'd have a better chance of survival. Meanwhile, we have the largest geographic range of any mammal. We inhabit all continents, remote islands, and dwell in habitats as diverse as rainforests, deserts, and tundra.

What Makes Us Human

Humans are generalists. The same went for species that survived the dinosaur-killing comet. They too rarely relied on single food sources. They were omnivorous beasts. Predators like alligators and snapping turtles that ate just about anything. We humans chomp down on literally thousands of plant and animal species. In different habitats we can be omnivores, herbivores, piscivores, or carnivores.

Foremost of human traits, however, is our ability to adapt. And the kind of adaptation we're talking about here is unlike any other species. It's not just adaptation through DNA—it's also adaptation through behavioral culture. Sure, we're animals. Sure, we're also mammals. But we are a very weird breed of mammal, for humans have science.

Instead of taking years, if not generations, to modify our genes by chance, humans use scientific intelligence. A culture of logical thought and rational tool-wielding to change our behavior in months, if not minutes. For example, ocean-going whales took many millions of years to evolve flippers and sharp teeth and sonar. But we humans took mere millennia to invent ships and fishhooks and sonar. Our recent experience with COVID-19 is another example. Viral genes evolve in a matter of days. But it takes merely seconds to tell another human to wash their hands. Cultural evolution outstrips viral evolution.

Not only is our scientific and cultural evolution faster than genetic evolution, it's also very different. Uniquely, and in the most profound ways, natural selection in humans has made us capable of intelligent scientific design. We don't just blindly adapt to our habitat. We knowingly remold it to our wants. More than any other creature, by far. Horses may have taken millennia to evolve the right teeth and develop complicated digestive systems to eat plants. Humans just domesticated plants, then felled forests for our crops. And cheetahs evolved twitch muscles in their legs to go at speeds fast enough to pursue their prey. Humans simply bred cattle and sheep that don't run too much.

Are Humans Born Preppers?

Are we humans so uniquely adaptable that we might survive the kind of mass extinction event that did in the dinosaurs? Imagine that we had ten year's notice of a comet strike. (In David Bowie's song "Five Years," a future Earth didn't have the luxury of ten years notice, as an impending apocalyptic disaster is about to destroy our planet in a mere *five* years, and the being who will save it is a bisexual alien rock star named Ziggy Stardust!) Given a decade of warning before a comet strike, we could prepper up. We humans could probably stockpile enough food to get through years of nuclear winter. Millions, if not billions, could survive this way. Longer-term crises, such as ice ages, might create conflict and population crashes, but civilizations are still likely to survive.

Yet, science and technology may have made us too clever for our own good. (It's a little like transgenically creating indominus rex and assuming that the future will be just dandy, thanks, with no possible repercussions.) In short, reshaping the world can often mean changing it for the worse. You just have to look at the catalog of nightmares dreamt up using science and technology: overpopulation and pandemics, nuclear weapons and waste, pollution and climate change. True, we've tried to alleviate these dangers with family planning and vaccines, nuclear treaties and regulations, pollution controls and solar power. It seems we've managed to avoid the worst-case scenarios of many science fiction dystopias. To date.

Our networked world, this global village of ours, has also invented ways to support fellow humans. For example, folk in the wealthy north often provide aid to poorer people in the global south, while at the same time creating the vulnerability of dependence in a crisis. Global travel, trade, and communications join up humans across the world. And yet criminal speculation on Wall Street trashes national economies elsewhere. Trouble in one country triggers murder on the other side of the world. And a virus from China, or wherever, threatens the future of millions.

Extinction at All?

Perhaps we should look to the future with guarded optimism. Take the recent COVID-19 pandemic, for example. English television screenwriter Charlie Brooker, most famous for his Netflix series *Black Mirror*, was pleasantly surprised about the initial human reaction to the pandemic:

> Because I've really always expected something like this to come along, I think maybe I'm not going through quite the level of psychological adjustment as some other people. If you look at what happens in classic dystopian fiction— where everyone turns on each other immediately—so far, that hasn't happened. It's not to say it won't. But I pivoted quite early to an optimistic view that this is terrible but, at the end of it, there's a possibility that we'll have the stomach to realign

society a little. Is this forcing our hands to address financial inequality and climate change? You hope that's the outcome, rather than that it makes psychotic strongman politicians more secure.

We might say the same for previous potential crises. Homo sapiens have survived over 250,000 years of ice ages, volcanic eruptions, and world wars. Maybe we'll survive another quarter of a million years, or even longer. The kind of pessimistic scenarios Charlie Brooker talks about depict disasters creating a breakdown of social order. Maybe even of civilization itself and the dawning of a dark, post-apocalyptic world. Yet, even in such worst-case scenarios, some humans survive, scavenging scant remains, Mad Max–style. Maybe even switching to farming, or hunter-gathering, though one finds it hard to imagine *that* scenario in the Mad Max franchise. But, in the end, it's not just a question of whether we can survive the next quarter of a million years. After all, the dinosaurs lived for over 150 million years of deep time.

INDEX